* SOCIAL IMPLICATIONS OF BIOLOGICAL EDUCATION *

✳ SOCIAL IMPLICATIONS OF BIOLOGICAL EDUCATION ✳ Arnold B. Grobman, Editor

THE DARWIN PRESS

PRINCETON, NEW JERSEY

This book was produced and designed under the direction of
John T. Westlake Publishing Services.
Typeface is IBM Baskerville with Univers.

FOREWORD

The National Association of Biology Teachers is pleased to make this volume available with the hope that it will be used to enrich biological education. Social studies teachers and students, as well as biology teachers, should profit from its contents.

The preparation of this volume was under the direction of the Social Implications of Biological Education Committee of the Association. The members of this committee were Mr. Leon Jordan, Dr. Paul Pearson, Mr. David Kraus, Dr. Garrett Hardin, Dr. L. Joe Berry, Dr. Gerald Scherba, and Dr. Arnold Grobman, Chairman. A special word of thanks is due to Dr. Grobman for compiling the manuscripts and editing the volume.

The work of the Committee on Social Implications of Biological Education came to fruition at the second annual NABT Convention at Philadelphia in 1969. The papers in this volume served as the major foci of the convention and many thanks are due to Dr. L. Joe Berry for arranging the program and to Miss Nancy Stees, the general chairman of the Philadelphia convention.

NABT appreciates the contributions of the main speakers, panelists, and NABT members who participated in the question--and-answer sessions following the major addresses.

Lastly, special thanks are due to Dr. Jerry Lightner, Executive Secretary of NABT, who helped make all of this possible.

Burton E. Voss
President, NABT, 1969

LIST OF CONTRIBUTORS

Leonard C. Blessing, *Millburn Senior High School, Millburn, New Jersey.*

E. Marie Boyle, *Penncrest High School, Lima, Pennsylvania.*

Lois DeBakey, *Baylor College of Medicine, Houston, Texas.*

Michael E. DeBakey, *Baylor College of Medicine, Houston, Texas.*

David Denker, *New York Medical College, New York, New York.*

V. G. Dethier, *Princeton University, Princeton, New Jersey.*

Arnold B. Grobman, *Rutgers University, New Brunswick, New Jersey.*

Garrett Hardin, *University of California, Santa Barbara, California.*

Ralph Hillman, *Temple University, Philadelphia, Pennsylvania.*

James C. King, *New York University School of Medicine, New York, New York.*

Haven Kolb, *Hereford High School, Parkton, Maryland.*

James V. McConnell, *University of Michigan, Ann Arbor, Michigan.*

Lawrence Mann, *Rutgers University, New Brunswich, New Jersey.*

Martin W. Schein, *West Virginia University, Morgantown, West Virginia.*

C. Richard Snyder, *Community College of Philadelphia, Philadelphia, Pennsylvania.*

Charles H. Southwick, *The Johns Hopkins University, Baltimore, Maryland.*

Burton E. Voss, *University of Michigan, Ann Arbor, Michigan.*

Bruce Wallace, *Cornell University, Ithaca, New York.*

Claude A. Welch, *Macalester College, St. Paul, Minnesota.*

Robert E. Yager, *University of Iowa, Iowa City, Iowa.*

CONTENTS

* SOCIAL IMPLICATIONS OF BIOLOGICAL EDUCATION *

CHAPTER ONE ✳ INTRODUCTION

by ARNOLD B. GROBMAN

Until quite recently, the dominant ethos in the teaching of science has been emphasis upon fact, with but passing reference to theories (especially those that might be controversial), and the avoidance at all costs of the social and ethical aspects of science and technology. Teachers of science, their students, and the professors responsible for the education of teachers were all in rather general agreement that science instruction had a circumscribed role with easily recognized boundaries.

This well-ordered scheme of things was abruptly challenged by a veritable revolution that erupted during the late 1950's. The implementation of this revolution in the philosophy of science instruction has involved a de-emphasis on the accumulation of facts and the substitution of a deep concern for the structure (the anatomy) of the sciences. A most significant aspect of the concern for structure is the encouragement in students of an exploratory attitude often referred to as an inquiry or discovery approach to science. Such an approach often encourages inquiries not only about the methods and nature of science but also about the social and moral implications of science.

Thus teachers who were trained for a teaching program focused primarily on the facts of science frequently were thrust with their students into penetrating considerations of the problems generated at the interface between science and society. In other words, science teachers who had been prepared for one kind of responsibility suddenly were expected to assume an entirely different kind of responsibility for which they had been offered no formal preparation.

Perhaps predictably, three generalized responses to this new challenge seem to have arisen among science teachers.

For some teachers, original training or personal inclination dictates their behavior; such teachers conduct their classes almost as if there had been no revolution in science education. Even if they use the newly designed, inquiry-oriented curricula, these teachers tend to respond to student questions by saying, "That's not part of our course," or "Please ask that question of your social studies teacher."

A second group of teachers views the revolution not only as a relaxation of previous sanctions but also as a license to project personal prejudices under the guise of free inquiry. Some of these teachers might condemn birth control techniques while others might extol them. Still other teachers in this group allow their personal political philosophies to strongly influence the directions of their classroom discussions of such topics as pollution control and pesticide restriction.

Teachers in a third group view the science education revolution as a most desirable improvement and are anxious to participate fully but are fearful that their own backgrounds are not sufficient for the balanced, exciting, and penetrating discussions they hope will occur in their classrooms. They are anxious to explore more deeply all aspects of questions that might arise in their classes.

Thus, in our rapidly changing circumstances, some teachers tend to avoid problems; some tend to be authoritarian in handling those problems; and still others are anxious to develop full and balanced discussions of those problems.

It is for life-science teachers in all three groups that the National Association of Biology Teachers (NABT) has prepared the present volume. It is hoped that they will find this book to be of some value in meeting the exciting challenge of inquiry-oriented science instruction and its relevance to social problems.

Not only biology teachers but many others may welcome thoughtful reviews of pressing social problems associated with advances in science. Such treatments, of course, cannot be balanced to perfection, for who among us is sage enough to determine what proper balance is? On the other hand, most observers find no difficulty in identifying an account that is so self-serving that it represents a strongly biased presentation. In preparing this volume, we have tried to come closer to the first pole than to the second but we do not claim complete success.

Our procedure was to invite distinguished biologists who have exhibited deep concern for significant social problems having a biological basis to address the national convention of the NABT in Philadelphia in October, 1969. All those whom we invited agreed to participate. Their papers were distributed, in advance of the convention, to panelists. These panelists were asked to prepare brief statements of their own which might support, contradict, extend, or restrict the major papers. The three members of each panel included at least one additional specialist in the field of the major speaker and one practicing school teacher who had exhibited interest in the topic under consideration by the panel. After each panel session, those in attendance at the NABT convention were encouraged to offer their own remarks and to address questions to the panel members. An attempt has been made to transcribe those question-and-answer sessions which often developed into quite lively discussions. No less than five hundred biology teachers attended each of these sessions.

Thus the next chapter, Chapter Two, is based on a paper by Drs. Michael and Lois DeBakey on the social implications of medicine, and includes the comments of the three panel members and a summary of the discussions of the question-and-answer period. Chapters Three, Four, and Five have a comparable organization and are based on papers on the social implications of behavior, by Dr. James McConnell; of genetics, by Dr. Bruce Wallace; and of population biology, by Dr. Garrett Hardin. Chapter Six departs from the pattern, because it is based on the banquet address by Dr. Claude Welch on evolution. This session did not include a panel discussion and a question-and-answer period.

At least three observations should be made about this volume so that no one may anticipate more than is intended.

The first and, perhaps, most obvious observation is that this volume is not intended to be a complete compendium of the social implications of biological education. It is selective and omits entirely some very important problems. For example, the whole vital area of the control of pollution and the redressing of our environmental balance is only hinted at in what follows; there is no chapter devoted to ecology itself which underlies a social and political problem of awesome dimensions. Many other gaps will be obvious to those searching for a complete coverage.

The second observation is that all the possible points of view for each topic are not necessarily presented. The goal was not to prepare an equal number of pro and con positions on each question so that teachers could spoonfeed students a fair "both sides of the question" lesson. Rather, the goal was to present a thoughtful collection of related statements for teachers to consider; these statements are intended to be stimulating and it is hoped that they will provoke discussion. They are not intended to be a collection of definitive statements on current problems.

The third observation is that many of the implications of biological education fall into two broad areas, one of which might be termed social and political, the other moral and ethical (or spiritual). Although the two are of course related, the present volume purposely attempts to explore problems in the first area only, and not in the second.

One further word about the genesis of this volume may be of interest. One of the few social problems that biology teachers of a generation ago did feel was appropriate for their consideration was instruction in conservation. In this the NABT has been an active participant, and in 1955, under the leadership of the late Dr. Richard Weaver of the University of Michigan, the Association produced the *Conservation Handbook,* which has been highly regarded. When his term of office as President of NABT was completed, Dr. Ted Andrews recommended that the Conservation Committee be reconstituted and broadened in its responsibilities.

Pursuant to that recommendation, the NABT established a Committee on the Social Implications of Biological Education which was responsible for the planning of the Philadelphia Symposium (referred to above) and the present volume. The members of that Committee, who were involved in the Committee's planning activities, included Dr. Paul G. Pearson, Rutgers University, New Brunswick, New Jersey; Dr. L. Joe Berry, Bryn Mawr College, Bryn Mawr, Pennsylvania; Dr. Garrett Hardin, University of California, Santa Barbara, California; Mr. Leon E. Jordan, Camelback High School, Phoenix, Arizona; Mr. David Kraus, Far Rockaway High School, Far Rockaway, New York.

and Dr. Gerald Scherba, California State College, San Bernardino, California; in addition to myself.

I am sure the Executive Secretary of the National Association of Biology Teachers would be pleased to transmit to any succeeding committee that the Association might establish all suggestions received from readers of this volume.

CHAPTER TWO ✳ MEDICINE

CHAPTER TWO ✳ MEDICINE ✳ *Social Implications of Medicine*
by MICHAEL E. DEBAKEY AND LOIS DEBAKEY

BENJAMIN DISRAELI assigned the highest social priority to health when he said: "The health of the people is really the foundation upon which all their happiness and all their powers as a State depend." It is no coincidence, therefore, that the healthiest people are the most ingenious, imaginative, and happiest, and that the healthiest nations are the strongest and most influential. An unhealthy society, on the other hand, is a defensively weak, economically unstable, morally feeble, and intellectually and culturally depressed society. The social implications of health and health care are therefore far-reaching and deserve everyone's serious consideration.

Health affects all our values—social, economic, cultural, political, psychologic, legislative, and many others. Our major social problems today—crime, poverty, malnutrition, urban crowding, overpopulation, illiteracy, civil discord, disease, alcoholism, drug addiction, accidents, and environmental pollution—are all intimately related to physical, mental, and emotional health. If we are to remove these threats to our well-being, we must all work toward this common goal. Health therefore becomes a social obligation, and the medium through which health is improved, medical research, becomes a social responsiblity.

From a practical standpoint, this responsibility yields dividends for all of us, since any weak element weakens the whole. Because the infirm cannot assume their share of responsibility for the production of food, shelter, and other human essentials, others must bear the additional burden. Restoring the infirm to health would convert them from nonproductive to productive consumers.

9

More than thirty years ago, Henry Sigerist, noted medical historian, wrote: "There is one lesson that can be derived from history . . . that the physician's position in society is never determined by the physician himself, but by the society he is serving." As intelligent, educated citizens, we are all morally bound to take an active interest in, and to help shape, the direction of medical science and the policies that concern human health. To do this intelligently, we must understand certain basic concepts and processes of science. We must keep informed of the organizational aspects of health in our community, state, and nation, and we must acquaint ourselves with new health problems, such as air and water pollution, as they arise, and take steps to eliminate these menaces.

That medical research leads to improved health services is not difficult to demonstrate. Physicians have traditionally translated the newest scientific discoveries into practical benefits for their patients, and these benefits have prompted further exploration, society continually expecting and demanding better medical service. New knowledge thus acts as a stimulus for further research, which, in turn, yields additional new knowledge and better health care. During the past half century, the life expectancy of Americans has been extended from about fifty years to more than seventy years, primarily as a result of the scientific conquest of major infectious diseases that were previously fatal. The discovery of methods of inducing anesthesia vastly broadened the scope and benefits of surgery, and the discovery of the roentgen ray by Roentgen, and of radium by the Curies, led to an entirely new approach to the diagnosis and treatment of disease. Discovery of the communicability of disease opened the field of preventive medicine and public health, and recognition of the relation of disease to lack of sanitation effected sanitation reform. It was medical research that led to the discovery of vaccines for diphtheria, whooping cough, tetanus, and poliomyelitis; drugs for control of tuberculosis; and antibiotics for treatment of pneumonia and other respiratory, gastrointestinal, and systemic infections.

In the field of cardiovascular diseases, more progress has been made during the past fifteen years than in all previous recorded history, largely the result of laboratory research. Heart disease, once a sentence of death or severe disability, is no longer hopeless. Within the past decade alone, the overall death rate from cardiovascular disease dropped 7 per cent among persons forty-five to sixty-four years of age. The death rate for hypertension, which affects about 19 million Americans, has been reduced by 40 percent, primarily as a result of the development of more effective drugs in the research laboratory. Impressive reductions have also occurred in the death rate for rheumatic heart disease (33 per cent) and stroke (22 per cent), particularly among persons below the age of sixty-five years.

In recent decades, incredible advances have taken place in the treatment of many grave forms of heart disease. The artificial heart-lung machine, a product of the research laboratory, is now used daily in operating rooms to support circulation of the blood during repair of a diseased heart or segment of the circulatory system. Operations devised for most forms of congenital heart disease now allow children and young adults to lead normal lives, with normal life expectancy. Aneurysms and occlusive lesions of the aorta and major arteries, formerly incurable, are now correctable surgically. The spectacular new procedures of organ transplantation and implantation of mechanical heart assistors originated in the research laboratory.

Researchers are collaborating on a computer model of the human circulatory system. By manipulating variables in this artificial circulatory system which cannot be manipulated in human subjects, they hope to study the effects on other parts of the system and thus to understand better some of the physiologic and biochemical factors in the human circulatory system. A machine to duplicate human breathing patterns is being developed to elucidate the mechanisms operating in respiratory diseases.

In the surgical research laboratories at Baylor College of Medicine, work has been in progress for some time on development of artificial "skin." Synthetic fibers, particularly Dacron, have been found useful for replacing severely burned tissue. Artificial elbows, knuckles, wrists, and other joints are constantly being improved, with remarkable advancement over previous stainless-steel prostheses.

In cancer, too, researchers have made tremendous strides. Almost a million and a half Americans who have had a major form of cancer are leading productive, happy lives. One of every three Americans survives cancer today, as compared with one of every four only a few years ago.

In psychiatry, advances have been revolutionary. Gone are the snake pits of former years. The advent of humane treatment has been accompanied by major contributions in effective drugs for anxiety and depression and in successful treatment of other forms of emotional and mental illnesses; these have allowed many persons to resume normal, productive lives in their communities. As a result, the number of patients in state mental hospitals has been reduced by more than one hundred thousand in the past decade. The savings in public expenditures have been considerable.

Engineering and technologic ingenuity has made it possible for physicians to detect illnesses through the aid of isotopes and ultrasonics, to operate with cryogenic devices, and to implant miniature pacemakers in the heart. Image-intensifier screens and cinema technics permit study of movements of the heart and blood vessels with minimal radiation; electrodes can be imbedded in tissues for automatic measurement of oxygen and carbon dioxide tension; and thermistor needles can be inserted for automatic measurement of temperature of tissues. Platinum electrode technics are being used to detect shunts between chambers of the heart, dye-dilution technics to determine valvular insufficiency and cardiac output, and memory loops to store abnormalities in rhythm in intensive care units. Direct high-voltage shock can be used to convert ventricular fibrillation to normal rhythm, and arteriography to identify coronary arterial occlusion. The electron microscope has advanced study of the behavior and fine structure of cells and molecules, and computers have facilitated diagnosis and treatment of many disorders.

New applications are being found for the laser, including therapeutic coagulation of detached retinas and destruction of certain chromosomes. Picturephones will soon not only place the patient in instant communication with the specialist but will deliver the patient's complete medical record, including electro-cardiogram, electroencephalogram, roentgenogram, and other clinical data from widely disparate geogrphic sites.   Many of these realities of medical science and technology were considered, only a quarter of a century ago, to be fanciful ideas of visionaries.

The social impact of these and other medical and technologic advances is wide-ranging, especially in such recent accomplishments as hormonal control of fertility and chromosomal identification, with possible prevention of birth defects. Most of the recent quasi-miracles in overcoming fatal diseases and restoring doomed patients to health and productivity can be traced directly and exclusively to medical research, and all have obvious social implications.

But with all this progress, science and society are not satisfied; both demand that we continue to combat yet-unconquered diseases and new health hazards as they arise. The past record of research is impressive, but we have much more to learn about disease and much more to do to allow man to achieve optimal health and fulfillment in this life. The American people are enjoying the best health they have ever known, but we are still far from our objective. We have not yet uncovered some very basic information about the human body and its health; we do not know enough, for example, about why the heart beats, why blood clots, why some cells grow wild, or why some babies are born deformed. The causes of atherosclerosis and arthritis, which account for more death and disability than all other diseases, remain unknown. To probe these and other unanswered questions, science requires the support of all our people.

An investment in science yields high dividends. A recent cost-effectiveness analysis of expenditures for the medical sciences projected remarkable savings that would accrue if funds for medical diagnosis and research were increased. In arthritis, for example, an expenditure of less than two-hundred dollars per person would extend by five years the income-producing lives of thirteen million patients. The total national saving would be 1½ billion dollars, for a benefit-cost ratio of thirty-eight to one; that is, for every dollar invested in improved diagnosis and control, thirty-eight dollars would accrue to our national economy. Where can we find a better investment than this? In cancer of the uterus, one of the most common and most fatal forms of this malignant disease, an investment of 119 million dollars would prevent thirty-four thousand deaths; for every dollar spent, nine dollars would accrue to the national economy. The list could be extended indefinitely to include elimination or suppression of venereal disease, vehicular accidents, and many other health hazards. From a purely economic standpoint, therefore, it behooves us to remove as much of our population as possible from the ranks of the disabled and handicapped, where they constitute a tax burden, and to place them among the productive and employable, where they can contribute as taxpayers.

But justifying medical research by citing cold figures of benefit-cost ratios is alien to the humanitarian ideals of a physician. Every human being should have a fair chance in the struggle for survival, and every patient should therefore be provided with the optimal health benefits possible by current skills and knowledge. Can we therefore put a price tag on human life? What price shall we assign to a drug that will arrest leukemia or prevent blindness in a child, or to an operation that will restore a cardiac invalid to a normal, useful life? As benevolent human beings, each of us has the responsibility of doing everything he can to help every person lead as comfortable, healthful, happy, and satisfying a life as possible.

We are proud of our affluence, our military superiority, our democratic ideals, and our professed humanitarianism, but let me cite a few health statistics that may mitigate this pride somewhat. Cardiovascular disease, our number-one killer, still claims more than a million lives a year—more than half our total annual deaths. Thousands still succumb annually to preventable and controllable diseases. According to a national health survey during the year ending June, 1962, 52 per cent of people in the labor force (71.3 million people) had one or more chronic conditions. The loss in man-hours and productivity is indefensible. Fifteen million Americans still suffer from heart disease, an equal number from rheumatic and arthritic diseases, 10 million from neurologic disorders, and nearly a million from cancer. If present rates continue, cancer will strike about 49 million Americans alive today. Forty thousand babies who die each year would live if we took steps to reduce our infant mortality, which is far higher than several other less affluent countries. One million pregnant women get no prenatal care at all, and of the 36 million persons who need dental care, only 25 per cent receive it. Every year 52 million Americans are victims of accidents, and about fifty thousand of these die. Suicide alone accounts for more than six thousand deaths. Mental illness, alcoholism, and drug addiction cause untold suffering not only to those afflicted, but to their families as well.

The economic loss to the nation from these and other causes of premature death and disability is staggering. Prevention of disease is far more economical than treatment.

Thus far we have discussed the support of medical research as a responsibility of society. But the social implications of medicine impose a reciprocal responsibility—that of medical scientists and physicians to society. Today more than ever, society's support of medical research depends on the ability of scientists to show the relevance of their work to community and national welfare. To maintain the respect, confidence, and trust of the public, medical scientists must make sure that the search for new knowledge is pursued in the best interests of humanity and within ethical and moral bounds. The investigator should never permit self-interest or passion for research to blunt his reverence for human life and the right of self-determination. He must find the critical balance required to satisfy society's demands for the advancement of knowledge while abiding by its strictures to protect the dignity, privacy, and freedom of its individual members, regardless of any laxity of principles in other spheres of life.

The true medical scientist feels genuine concern for the patient as a human being, rather than simply as a subject of cold professional interest. This concern is important, for every patient must feel confident, when he entrusts himself to the care of a physician, that the physician will not conduct a therapeutic experiment on him until adequate scientific evidence is available to warrant it—and then only when the potential benefits outweigh the risks. At a time when all of us are acutely aware of a re-examination of ethical, moral, and social values, and when these values have impinged heavily on medical research, it is incumbent on physicians and medical scientists to protect the traditionally high ideals, objectives, and integrity of the profession.

Medical ethical criteria require the most sober deliberation and the most prudent consideration by the physician-scientist when he applies new scientific knowledge at the bedside. If a technic is highly experimental, for example, he should not attempt human application until he has adequate evidence of its safety and potential efficacy, usually in the form of extensive controlled experiments and observations in lower animals; he should anticipate and safeguard against additional risks peculiar to human application; he should obtain the informed, and totally voluntary, consent of the patient or his guardian after adequate explanation of the investigative nature and uncertainties of the procedure; and he should, ultimately and above all, assure himself that if he or a loved one were the patient, he would sanction the proposed procedure. When the decision is overly onerous, he should seek the counsel of informed peers, never depending upon himself as the highest or sole arbiter.

The social implications of such medical advances as contraceptive pills, cardiac transplantation, and the artificial heart are far-reaching and encompass not only moral and ethical issues, but legal, theologic, psychologic, and others. It is entirely proper that they be thoroughly examined, free of emotionalism and of the meaningless clichés, trivia, and illogicalities that cloud rational judgment.

The physician has traditionally been confronted with critical moral decisions, and has generally found the wisdom to make appropriate judgments, whether they involve a family with religious objections to blood transfusion or relatives pleading for the physician to discontinue active supportive measures for a patient terminally ill with an acutely painful disease or irreparable brain damage. For particularly weighty decisions, the physician normally consults not only his colleagues, but also the patient's family and religious counselors. This process has a social and legal precedent in the jury system, which provides wide representation of opinion and knowledge, and apportions responsibility for the decision. The physician is also guided by public opinion, for society has the prerogative of accepting or rejecting the conduct of any of its members. The medical profession is expected to be particularly sensitive to social reaction, having been granted special sanction and invested with special trust.

By tradition and by training, the physician's instinct is to prolong life and to restore physical and mental health; toward this end, he is expected to concentrate all his energies, scientific knowledge, and skills. That the life entrusted to his care is worthy of being saved is a necessary presumption, and he is not concerned with the patient's achievements before or after his medical ministrations. Nor does he proffer gradations of medical treatment according to the social worth of a patient; the convict is due the same quality of care as the brilliant, productive scientist, insofar as facilities and professional manpower are accessible to each.

The first human cardiac transplantation brought a flood of knotty questions, not only about whether the recipients of the limited supply of donor organs were to be judged on the basis of material or social worth, but also about when it was permissible, on clinical grounds, to remove a functioning heart from one body for transplantation to another. A more precise definition of death became necessary because, with present scientific knowledge, viability of an isolated heart outside the human body can be maintained for only about one-half hour. When scientific evidence indicates that a patient has suffered irreparable brain damage, but that his heart and other organs can continue to function mechanically, the physician must decide whether to allow two patients to die, one with irreparable brain damage and the other with incurable heart disease, or to try to save one life by transplanting a healthy, normally functioning heart from the decerebrate patient to the one who is mentally alert and has a chance for survival.

In the early moments after a tragic accident, the family of a victim is naturally eager that everything possible be done—regardless of how difficult, impractical, or costly—to save the patient's life; but once they have witnessed the greater tragedy of a lingering, insensate body which no human effort can restore to its former mental and physical state, most will accept the inevitability of human death, and some are eager to help the patient, in death, grant someone else another chance to recover a healthy life.

The controversy over moral and ethical issues in organ transplantation emerged partly from the conflict between the rights of the individual and the rights of society. Although the physician's first obligation is to his patient, he also has a responsibility to society. He must therefore balance the rights of the individual against the rights of society in certain kinds of human experimentation. Instead of taking either extreme view—that the individual should subserve society or society the individual—the physician tries to minister to both, and in no case to do disservice to either. *Primum non nocere*—first not to harm—remains the physician's obligation today, two centuries after being enunciated by Celsus. This precept holds for any new clinical definition of death.

Some have suggested placing legal restrictions on medical experimentation but obedience to an ethical code should not have to be exacted by rigid, formal laws or injunctions; it should, instead, be prompted by integrity, humanitarianism, and benevolence—qualities that every physician should possess. Ethical codes have been formulated as guides to proper action, but the physician's conscience remains the ultimate guide, as in all human conduct. These medical ethical codes, which are based on compassion and integrity, have served many generations of physicians nobly. When, however, individual members of any group violate basic social principles by failing to exercise restraint, prudence, and discretion, rules will invariably be imposed from without. Ethical decisions in medical science depend finally on the wisdom, integrity, and self-imposed restraints of the scientist and his peers. A simple personal credo based on general ethical norms and on love and reverence for humanity has no equivalent for moral guidance. The Golden Rule is an excellent guide for any physician-scientist to follow: "Whatsoever ye would that men should do to you, do ye even so to them."

In the future, ethical, sociologic, economic, philosophic, legal, and many other issues will bear more and more heavily on the direction of medical research and on the character of health care. The future status of medicine, like its past and present, will grow out of the complex interplay of multiple forces, and will continue to have significant social implications. Paramount among forces that have influenced the evolutionary process in the past have been the expansion of scientific knowledge through medical research and the application of this knowledge through ingenious technologic developments, both of which have revolutionized health services and have effected vast social changes.

Unlike his predecessors, who accepted the moment of life or death as predestined, modern man can exercise some free will in both. By means of artificial insemination, for example, he can control events in the creation of human life. Science has, in fact, brought us to the threshold of the secret of life. To some extent too, man can control the time of death, sometimes preventing it by previously impossible or unknown heroic measures, such as cardiac massage, mouth-to-mouth resuscitation, electrical defibrillation, renal dialysis, organ transplantation, and mechanical cardiac assistance. The cost of some of these is considerable, but these and even more remarkable lifesaving procedures will one day be available to all people.

Future medical science undoubtedly holds discoveries that the contemporary mind would find even more fantastic than those mentioned here. The resulting changes in the mode of practicing medicine will be wide-ranging. Anesthesia induced by injection of pharmacologic agents, for example, may be replaced by a form of suspended animation or hibernation that will permit temporary cessation of vital functions while corrective treatment is being applied. We can look forward to the time when enough will be known about prevention of arteriosclerosis to remove this major cause of death. Discovery of the origin and prevention of cancer should eradicate this baneful disease. The cause of mental illness will undoubtedly be uncovered, and methods of preventing birth defects will eliminate the dreaded risk of having a deformed baby. But these discoveries are contingent upon an informed public willing to support medical research.

Society must weigh the advantages and disadvantages of such investments against its other needs. In such deliberations, it should recognize the broad social implications of medical research and its offspring, improved health. For health, the goal of all medical research, is essential not only to the integrity of a nation as a whole—its national defense, economic stability, and intellectual, cultural, and scientific development—but also to the fulfillment of its individual members.

In closing, we would like to urge all teachers of science not only to convey to your students the basic concepts, the philosophy, and the methods of science, but also to keep your students informed of the vast social implications of medical science, so that each student may recognize his responsibility in helping to shape public policy in this field. Your students who choose a career in medical science will have an opportunity to contribute directly to the betterment of human health. And *all* of them can help achieve this goal, first by maintaining their own health and then by extending their concern to the community, nation, and world. Public investment in medical research will yield unimagined dividends to all of us.

CHAPTER TWO ✳ MEDICINE ✳ *Some Social Problems Caused by Scientific Advances* by LEONARD C. BLESSING

Without question, one of the areas where science has made the greatest and most obvious contributions to public advancement is in the field of medicine. The understanding of the relationship of the mosquito to malaria, the use of insulin in controlling diabetes, the reduction of typhoid fever incidence, and the development of the vaccine for polio are just a very few of the important gains for the public.

But in the mad rush to make the world a better place in which to live, science has made some drastic mistakes. The use of radioactive materials on watch dials led to disfiguration and death of many women who worked on the production of these watches. Many insecticides have side effects that seem more destructive than the expected benefits. Fertilizers have increased the yield of our crops spectacularly but the resulting wash-out has introduced algal blooms in rivers and lakes to uncontrollable levels. It has recently been discovered that strontium 90 from earlier atomic explosions may be responsible for the death of large numbers of infants in their first year of life. And the cyclamates that we have been consuming may be responsible for the introduction of genetic defects into our population.

In recent years the general public has become somewhat disenchanted with science. Of course many persons are willing to accept their personal gains from modern technology, such as air-conditioned automobiles, TV sets and frozen foods. But what is the response to other gains of science, such as the fluoridation of water or information about the health implications of smoking?

Let us even test ourselves for a moment. We are supposed to be teachers of science and developers of scientific thought. How would you answer this question? We have learned how to detect whether a child being carried by a pregnant woman will be a Mongoloid. To allow such a defective child to enter the world would certainly introduce a strain on the parents and an expense to society. What is your decision on abortion?

The advances in the field of organ transplants has proceeded at a remarkable rate. Kidney transplants are now established procedures; heart transplants seem to have had considerable success; and liver, lung and pancreas transplants have hardly worked at all. The field of bone marrow transplants seems to offer some high hopes in the fight against leukemia. These are surely new and exciting fields of interest.

Another medical problem of considerable importance is the preservation of unfavorable genes in the human gene pool. I do not mean to be crass and unsympathetic; if one of your loved ones is involved, you certainly want him to continue to live. But in our less emotional moments, can we not think of the possible implications to our race by keeping alive so many people who will reproduce and pass on defective genes to future generations? I do not suggest a lessening of our medical improvements but I think we should face the problem head on.

I believe in improving the yield of crops, in fertilizers and insecticides, in going to the moon, in genetic improvement, and in cellular chemistry, but as educators of biology we must introduce into our science courses concerns about the social problems of today. Our modern courses of Chem. Study, PSSC and BSCS are intensely interesting courses to the future scientists, but they do not reach the vast majority of students who want to know what science can do for, and how scientists feel about, the morality of birth control, pollution of the land and waters, construction of atomic energy plants on rivers, wanton destruction of natural resources, and man's plight in our cities.

CHAPTER TWO ✻ MEDICINE ✻ *Keeping Students Informed of the Social Implications of Medicine* by ROBERT E. YAGER

The Drs. DeBakey have urged teachers of science to "keep our students informed" about the social implications of medicine. I agree that this is a most important function of science instruction for all educated students, whether at the elementary, secondary, or college level. Public concern for and involvement with science and medical decisions means the necessity of an informed public. Science teachers at all academic levels are doing all too little to create such an informed public. We need courses, learning sequences, institutes, and new pre-service programs for teachers if such changes are to be realized. We have done much in the new elementary, secondary, and college programs to bring most students to the forefront of scientific research. We've emphasized the basic concepts of science, the philosophy of science, the methods of science. We have done little at any level to emphasize the reciprocal effects of scientific research (medical and otherwise) and society about which the Drs. DeBakey have commented.

Perhaps we cannot do this task alone. Perhaps the departmentalization that characterizes secondary and college faculties precludes accomplishing such a task. We need more dialogue with teachers from all academic areas. We can then consider the problems of man from a variety of perspectives. We are seeing such moves for interdisciplinary courses in all too few secondary schools and colleges.

We need to be educating our students for change—not in specific content. No matter what characterizes our courses in terms of content, it will not be the important content for our students in adult life. We need not only to note changes in science and society but we need to consider possible changes in the future and learn possible means of dealing with them.

The DeBakey paper has predicted many medical breakthroughs that are on the horizon and we need to prepare our students who are in our schools today. We need Dr. DeBakey and others from the field of medicine to help our schools prepare courses that will lead to an informed public in future years. For example, college scientists, high school teachers, and other educators have collaborated to produce the exciting Biological Sciences Curriculum Study courses. We need help from medical researchers and others in the medical professions if we are to consider adequately the social implications of medicine in our schools. We need to have the best estimates of what the future holds for us; we need to know the problems, the directions, and the likely speed of new advances; we need to feel the common societal bonds with colleagues in the medical sciences if teachers of science are to be prepared for the task of even more curriculum reform. We need persons who will make predictions, consider alternatives, and help supply appropriate ideas for our schools.

Scientists like Dr. Herman Kahn of Hudson Institute are all too few. As you may recall, Dr. Kahn has predicted extensive use of robots "slaved" to humans; complete control of marginal changes in heredity; chemical control of intelligence; and some direct control of thought processes. We do not need more bizarre predictions than Dr. Kahn's. But we do need sources of informed thought as a basis for discussion.

Another problem of social implication to which the DeBakey paper has alluded is that of societal control through provision of funds for further research. We can all call for continued (or added) public support in order that many of the remaining medical and health problems can be attacked and solved. Joshua Lederberg, in a recent column for the *Washington Post,* has identified another basic problem in this age of governmental support and control of research. He is critical of and concerned with a society which is involved with "buying given projects and specified results." He questions the future of a research which is not based upon independent initiative in which the investigator is his own critic. Would a public informed of the dangers of such approaches allow the politicians and others in government to make decisions regarding the direction of scientific research?

Robert S. Morrison has recently asked, "Have our best doctors become so preoccupied with the wonders of their technology that they have become indifferent to the plight of large numbers of people who suffer from conditions just as fatal but much less interesting? Even the most earnest advocates of increased research in heart disease, cancer, and stroke must be a little embarrassed by the fact that the United States, which used to be a world leader in reducing infant mortality rate, has now fallen to fifteenth place." The DeBakey paper mentions the U.S. mortality rates in babies as one of a series of areas where "health statistics" do not paint a picture in which we can take pride. However, the decisions concerning which problems we attack—which ones we support —what changes we really desire—are questions for society as a whole. The problem of this kind of decision making is one we need to consider in schools, for schools should reflect our culture as a whole. An interaction among all the components of society should characterize modern classroom discussions. What are we really doing to prepare our students for the real world? Is the main thrust of our courses preparation for change itself? Or, are we still concerned with the mere identification and learning of the "basic concepts" of biology, devoid of any social implications? The Drs. DeBakey have made many profound statements in their address. We all need to consider carefully the questions they raise and the solutions they urge. I, for one, hope that the DeBakeys and other leaders in medicine will help with both!

CHAPTER TWO ✳ MEDICINE ✳ *Comments on the DeBakey Paper*
by DAVID DENKER

What emerges from the DeBakeys' paper is that we are all accustomed to thinking of our time as the age of science; we glory in the description. There is an intoxication about the speed of our progress. These are visibly times of greatness in both physical and theoretical achievement. In all the sciences, and nowhere more clearly so than in the medical sciences, the inconceivable is conceived, the unthinkable thought, the undoable done. The human dimension has never before been so enormous, and the catalogue of human resources so nearly infinite. After thousands of years of bare subsistence, the human race is on the eve of abundance for everybody.

If the pursuit of happiness is so nearly at the point of capture, why are we so deeply troubled? A sense of crisis is discernible on every hand; we feel it in ourselves, we hear it on the lips of friends and strangers. We live uneasily with ourselves even as we contemplate the promise of ease for everybody.

The scientist shares the human awareness of a crisis at hand. He is the principal architect of the accelerating changes which characterize our progress. More than most of his fellow men, the scientist is anxiously aware of the growing discrepancy between the achievements of his research and technology and their application to human life. Society is clearly failing to make a satisfactory use of its own new resources. There is a widening disparity between what actually is accomplished and what could be done. In medicine that discrepancy has created an "expectation gap," a sense of the insupportable difference between what a man or woman can reasonably aspire to these days and what the man or woman actually does receive.

The crisis of today is not one of knowledge. It is a crisis of responsibility and of social philosophy, a crisis stigmatized by inadequacy and unavailability. What we have is in short supply, and too few of us can get at it. We are crowding ourselves away from our own banquet table. A world confronted by an exploding population is further confounded by a second crisis of scientists in shortage. Central to this scientist shortage is the massive doctor shortage. Item: Today, in this nation of affluence, there is a great and growing doctor deficit. Already alarming in scope, it threatens to reach disastrous proportions. In the United States, the ratio of physicians per 100,000 population was 132.9 in 1960 and the USPHS projects 130.5 for 1970 and 125.9 for 1975.

Item: Thus far, I have mentioned the doctor deficit generally. Now, let us examine the situation in our hospitals. In September, 1965, about 13,000 internships were offered. Of these, 7,309 were filled by the graduates of United States medical schools, 3,284 (almost one out of four) were vacant, and 2,361 were filled by foreign graduates. In the same year about 39,000 residencies were offered of which 22,765 were filled by United States and Canadian graduates, 9,000 by foreign graduates and 7,000 (almost one out of three) remained vacant.

To examine the significance of these figures, let us compare college and medical school enrollment. Between 1959 and 1965 there was a 22.3 per cent increase in college enrollment and a less than 7 per cent increase in medical school enrollment. In the next ten years a future increase of about 65 per cent is anticipated in college enrollment. Thus, from 1959-60 to 1974-75 the increase in the United States college enrollment is projected at 100 per cent, whereas even if we succeeded in completing a total of twenty new medical schools in the United States and Canada, the most optimistic projection in medical school graduates for both countries is an increase of less than 30 per cent.

Item: The poorest nations, those in greatest need, are losing their already woefully inadequate number of doctors to wealthier lands, including our own. India, with a ratio of one doctor to 5,800 people is today being drained by the United Kingdom with one physician for 840. We, in turn, drain the United Kingdom. Where we should be proudly exporting our medical skills to Latin America, Africa, and Asia, we are instead siphoning their sorely needed doctors into our medical economy. Consider the sad consequences of the medical "brain drain." The United States drains England, and England drains India, Pakistan and other Commonwealth countries. In France there are as many Togolese doctors teaching and practicing as there are in Togo. And in Iran the situation is even worse. There are more American-trained Iranian doctors in New York than in Iran.

This then is the disquieting question: how do we deal with a shortage of doctors that has reached alarming proportions at home and abroad? The rational answer is that all medical institutions have the obligation to increase the total number of doctors at least 100 per cent and to make available much higher proportions of medical personnel to countries which sorely lack doctors. My own experience suggests to me that there is no substitute for excellence, but the quality of our performance and our facilities must advance at a pace to match the serious manpower problem. The doubling of our enrollment is essential in this process.

The medical community of this nation properly should pledge itself to the health of man—not one man, not one race of man, and not one nation of man, but to all mankind.

CHAPTER TWO✳Discussion

QUESTION—*Aren't doctors more likely to devote attention to certain persons as opposed to others?*

DR. DEBAKEY: I think that the dedicated physician would exercise the same kind of care with an impoverished patient as with a wealthy or prominent patient.

QUESTION—*Do you subscribe to the AMA philosophy?*

DR. DENKER: I do think that there are healthy signs on the landscape. The AMA has recognized the problem I have tried to define here; so has the Association of American Colleges. The task, of course, is to get the medical schools to increase radically the size of the entering class.

QUESTION—*What can medical schools do to increase the supply of physicians?*

DR. DENKER: It is probable that by 1975 one out of every five or six people will be engaged in health-service industries and professions. By the turn of the century, however, as many as one out of each three or four employed persons could be engaged in health-service industries and professions. The implications of such developments are staggering, not only in terms of the technical manpower needs these demands will create, but also in terms of the effect on our political structure. You will readily understand why we must immediately address ourselves to all aspects of our doctor deficit and our scientist shortage. Medical care is today not a privilege but a public right.

The disquieting question: each medical school in the nation will have to reorient the curriculum to revise its philosophies to contend with a changing world.

QUESTION—*Is there not a real role in which paramedical people can help physicians reach more people?*

DR. DENKER: There is indeed. The recent example of a medical school that has recruited Navy corpsmen is a sound idea. We must devise new and better ways of training paramedical personnel to relieve the physician of time-consuming chores and to accommodate the increasing manpower needs of the advancing technologies of medicine.

QUESTION—*Reference has been made to increasing the life span and preventing birth defects and infant mortality. The focus of my question is, what is the impact of the population problem?*

DR. DEBAKEY: The solution to the population problem is the responsibility of all of society and not the medical profession alone. Controlling the birth rate is a more humane solution to the population problem than withholding attempts to prevent birth defects, lower the infant mortality rate, or prolong life in those who might, with treatment, have additional happy, productive years on earth.

QUESTION—*Do you believe that government intervention is necessary to create more medical schools?*

DR. DENKER: Well, I believe in benevolent intrusion of government on all levels—local, state, and federal. This does not mean that the universities, the foundations and the corporate sector can duck their responsibilities. We observe, for example, that as the federal government moves in the foundations tend to move out. I also worry a little about the way the research mission can be distorted by too great reliance on national agencies for support. Here, however, is the irony—unless the federal government assumes a larger share of the burden a number of medical schools will be in serious financial trouble. The idea of an academic depression is no longer academic.

QUESTION—*Although we certainly do need more doctors, is it not clear that increasing the number of people in medical schools might lead to lower standards?*

DR. DENKER: I have tried to make clear that my own experience suggests to me that there is no substitute for excellence, but that the quality of our performance must advance at a pace to match the serious manpower shortage. What is needed is the shaping of a few approaches to meet new challenges. And I am sure that this can be done without sacrificing quality. My large fear suggests that failure by education to initiate those approaches may result in political pressure from local, state, and federal authority.

QUESTION—*Please give us a little more detail about the institutes.*

DR. YAGER: I don't know whether I've thought out a whole plan to propose here. My big plea is for a place where we can talk to the Drs. DeBakey and others who have important views as to the future—problems, research, and breakthroughs. We have had institutes devoted to the training of teachers and writers in connection with the various curriculum projects. My plea is simply to involve other people who are knowledgeable about where we are going, where research is going, where medicine is going, where some of these problems are. We need to get together and talk about these issues in order that as a community of teachers we can be better informed. We would be in a better position to plan our courses and new directions for our science courses. It would make no difference to me whether such sessions were conducted on a university campus or whether they were planned by a group of teachers, by a group of college instructors, by a group of medical educators, by specialists concerned with population problems, or by any other scholars. My plea is simply that we need to talk with all kinds of informed people if we are going to be knowledgeable and if we are going to effect certain changes and do certain things that Dr. DeBakey has urged in the schools. What goes on in our schools needs to reflect our total culture. What goes on is not merely the result of teacher opinion and background from post-college experiences.

QUESTION—*Reference has been made to the shortage of doctors. Isn't there some way computers could help expedite some of the routine aspects of diagnosis?*

DR. DENKER: Yes, the use of computers is what I had in mind when I indicated the need to revise our philosophies and policies. Another thing: anxiety and frustration are enemies of a sound and developing mind. A medical student from an underdeveloped country who is trained in the United States in diagnostic techniques will not be prepared for the conditions he finds when he is sent back home. We must establish programs abroad to reorient medical students, particularly in the less developed countries. We should also work closely with colleges abroad to produce more medical students in their regions.

QUESTION—*We're apparently giving our medical students good technical training—are we apprising them of the social problems they will be facing in practice?*

DR. DENKER: I think Dean Grobman understands this problem better than all of us here. Believe me, it is easier to pave over a cemetery than it is to change the curriculum. The task, I suppose, is to humanize the scientist and to simonize the humanist.

QUESTION—*What can teachers do outside the classroom to be effective in solving some of the environmental problems?*

DR. GROBMAN: The simple answer is to be as good and effective citizens as you possibly can—take advantage of all opportunities to encourage your legislators at all levels to become concerned and join with other citizens in pushing for things you think in our best interests. I have no special formula other than those general remarks of being good citizens and using your knowledge as effectively as possible. I would also urge that when you use your knowledge, you be reasonably sure that you are accurate. Sometimes, in our enthusiasm, we say things we can't always defend and then our whole arguments are very badly weakened.

QUESTION—*Reference is made to remarks that the community should be more involved in the planning of curricula for our schools. In terms of institutes, shouldn't there be more provision for teachers to be involved in an institute program designed for their own education?*

DR. YAGER: I couldn't agree more that teachers should be more involved. I think that many institutes have been too strongly centered around a college faculty, i.e., a college microbiology department, and I think these people have often been very honest about what they thought would be needed. However, in my opinion, too much effort has been devoted to the teacher's mastery of more content information in many programs in the past. In a sense, this is what I am talking about in terms of medicine: I think we need to be deeply concerned with some of the big problems associated with medicine. We need to talk with the people involved, and the needs of the teachers should be considered. This does not mean that we would train teachers as paramedics. In my opinion an institute is not a college experience when a given set of college professors "lecture" to a group of teachers who are enrolled for credits and grades. An institute should be a place where intellectual exchanges can occur among participants and staff.

QUESTION—*Part of your remarks indicated that we have the highest medical standards in the world and another part says that we do not have the highest insofar as infant mortality is concerned. How do you reconcile the two statements? What should be done?*

DR. DEBAKEY: My statement was that the American people are enjoying the best health *they* have ever enjoyed—not the best health that can be attained. In answer to your second question, there are several ways health care can be improved. One is to take Dr. Denker's advice and increase the enrollment in medical and allied health schools so that we will have more physicians and other health professionals to care for our growing population, including the underprivileged segments of our society. This will require additional funds, of course. We also need to strengthen our support of medical research, for it is through research that scientific discoveries are made and are eventually applied by the physician at the bedside of the patient. The retrenchment in federal funds for medical education and research will seriously affect the health care of our people. If all of us wrote to our congressmen and expressed our discontent with the new health appropriations, we might be able to have some of the cutbacks restored.

CHAPTER THREE ✻ BEHAVIOR

CHAPTER THREE ✻ BEHAVIOR ✻ *Ethics and Behaviorism* by
JAMES V. McCONNELL

Let me begin with what may seem a rather wild-eyed statement
that not all will be willing to accept: The time has come when,
theoretically at least, we can gain such complete control over a
person's behavior that we should be able to change him from
whatever he is to whatever we think he ought to be. I make this
statement with the kind of implicit faith that probably prompted
certain physicists to tell President Roosevelt in 1940 that they
could build an atomic bomb. These physicists had a great deal of
experimental data, all of which suggested that a bomb could be
built; they couldn't prove they were right, of course, until the
American government provided several billion dollars to bring the
Manhattan Project to fruition. Similarly, I can't give actual demon-
strations that people can be changed as readily and as drastically as
my opening statement suggests, but I do think that the data are
firmly on my side, and we must keep this perhaps revolutionary-
sounding thought firmly in mind if the remainder of this article is
to make much sense.

According to the newspapers and magazines, these are truly
revolutionary times. There is the revolution in education that
frightens college presidents, exhilarates the students, and leaves
most faculty members vaguely confused. There is the sexual revolution that has brought with it the open sale of such underground
classics as *Fanny Hill* and *Tropic of Cancer,* not to mention the
public exposure of more square inches of human flesh than our
grandmothers would have deemed physically possible. And there
are the political revolutions, both here and abroad, in which many
minority groups are demanding, for the first time in history, their
fair share of the political plums (if not the entire plum tree itself).

These are the stories that catch the headlines, for most Americans rather like reading about such titillating topics as sex and violence. But frankly, I believe that the sexual, educational, and political upheavals that we see around us are but different aspects of a larger change that involves how man views himself. I call it the Behavioral Revolution.

Most of the really important revolutions that have occurred in the past few centuries have to do with the way man perceives either himself, his neighbors, or the physical world around him. The American and the French political revolutions, for instance, can be seen as the destruction of the aristocratic view towards the human race. Until the 1700's, the majority view in the Western world was that some people were simply better than others, that God in his infinite wisdom had set the aristocrats above the commoners, and that this social-political disparity was right and just and moral and ethical and, perhaps above all else, a fact of nature. The commoner usually saw himself as being inferior to the aristocrat—or at least most commoners unsullied by democratic propaganda probably saw themselves in this light. The Magna Carta, Martin Luther's manifesto nailed to the cathedral door in Worms, the American Declaration of Independence, Lincoln's proclamation freeing the slaves, the constitutional amendment giving women the right to vote—all these documents are part of the revolutionary process that leads from the divine right of kings and popes to the type of complete political and social equality we are striving for here in America. But the most important aspect of this political evolution lies not in the fact that the aristocracy is no longer a highly privileged class nor even in the fact that today everyone can (theoretically) vote. No, to my mind, the crucial change has been in the way that man perceives himself in relation to his peers. For true political and social equality will come only when all of us actually *perceive* this equality as being the true state of affairs.

I like to think that the first real revolution in Western civilization began about 1600 when astronomy and physics emerged as scientific disciplines. Prior to this date, Western man's view of the world around him had been influenced almost totally by his interpretation of the creation of the universe described in the Bible. God had created man but a little lower than the angels and and had given him dominion over all the rest of the physical world. Obviously then the earth had to be at the center of the galaxy and the sun had to revolve around the earth. When Galileo and Bruno insisted that the earth moved around the sun, they were attacking not only man's view of the physical world, but also his perception of himself as being in the center of creation. No wonder that the authorities burned poor Bruno for heresy, and that the Catholic Church has still not officially forgiven Galileo.

Once man was willing to look at the physical world as a product of natural forces, he could study it objectively without the perceptual biases and emotional hang-ups that prevented earlier scholars from making the discoveries they otherwise might have. And, of course, once man had a new theoretical outlook on the physical world, he was freed to develop a technology based on that theory, a technology that has, of course, profoundly changed our life space in the past three centuries. The average man, I'm sure, couldn't care less about which theory was correct in any absolute sense, but the average man is intensely interested in what technology has to offer. The scientific view of the universe es- poused by Galileo and Bruno won out not because it was a more accurate representation of the world, but because it led to bigger and better machines, to a more comfortable and richer life than did the metaphysical, theosophical view of the universe that preceded it. Man is above all else a practical animal—he loves his pleasures and hates his pains and he who offers him more of the good things of life and fewer of the bad things will win his allegiance. The physicists and chemists for centuries have been at the top of the pecking order among scientists; they got there simply because the technology that sprang from their work was so much more powerful than that attached to the biological and social sciences. Scientific theories, then, are ultimately accepted or rejected for very practical reasons.

It took mankind thousands of years to learn to be scientific about the behavior of the physical objects in the world around him. For most of recorded history, the human race has endowed sticks and stones and waterfalls and thunderstorms with "spirits" that made these inanimate objects "behave." In the last few hundred years or so, we've come slowly to realize that it's rather impractical to explain the actions of stones and thunderstorms in terms of spirits inside them, for praying to spirits simply doesn't give us the great control of the physical world that a scientific technology gives us. The physicist and the chemist then are free to be objective about the things they study. No one insists that a physicist love his electrons and treat them gently and humanely, nor do we insist that a chemist be particularly nice to oxygen molecules and help them develop their proper electrical potential because they've grown up in a deprived environment. The ethical constraints that we place upon physical scientists are aimed at controlling the technology that flows from their work, not at controlling the objectivity with which they view the physical world.

The biological revolution didn't really get off the ground until the nineteenth century, when men like Pasteur showed that looking at bodily processes objectively could yield immediate and immense practical rewards. Before that time, medicine was more an art than a science, mostly because early physicians refused to accept the theory that man's body was little more than a biological machine. Until man was ready to admit that scientific laws governed the actions of the various parts of his body, medicine simply could not become a science. And, of course, medicine could not become an *experimental* science until man was ready to admit that controlled laboratory experiments might yield valid knowledge about the functioning of his body. Most medical doctors are technologists rather than scientists, but the technology they apply is, for the most part, firmly based on scientific findings.

As late as the 1800's, most people in the Western world believed that physical illness was almost always a reminder from God that the sick person had fallen from a state of grace. The Bible, after all, is full of instances of plagues and poxes that God visited on his children because they had sinned or, as in the case of Job, merely to test their faith. The biologists insisted that one didn't need to postulate divine punishment in order to explain sickness and death, that all one needed to do was to take a new and radical view towards the functioning of the body. The biological viewpoint is all very logical and objective, but I rather doubt that it would have won the day had it not given birth to the wonders of modern medical technology. In the long run, the faith healers lost out because surgery is demonstrably a more effective way of curing appendicitis than is the laying on of hands.

The rise of medical technology points up a very crucial issue as far as ethics and morality are concerned. A hundred years ago, surgery of any kind was considered immoral by many people because they perceived the body as being God's temple and the Bible said it shouldn't be desecrated. The study of anatomy was hindered for centuries by this view of man's biological functioning, for until recently it was simply not ethical for a surgeon to cut open human flesh. But when over a period of a hundred or more years, the average man came to see that surgery saved lives, then autopsies and operations became morally acceptable.

Ethical considerations almost always yield to (and perhaps even follow from) practical changes in the world. The Old Testament proscription against eating pork, for example, might well have sprung from the fact that the flesh of swine often contained nematode worms that cause trichinosis. I wonder how many young Jews consider this dietary injunction a real ethical problem, now that modern sanitation has made trichinosis relatively rare in the Western world? Likewise, I think we can predict that, despite the Pope's adamancy on the subject, most young Catholics do not see birth control as being the pressing moral issue that it was, say, a generation ago. It is quite clear to most astute observers that the Catholic Church is going to do an about-face on the "pill" because, in the long run, practicality determines what is ethical and what isn't, which is to say that practicality determines how we see ourselves and the world around us.

Most people in the Western world today believe that man is the master of his fate, that he consciously and deliberately chooses what he is going to do and when he is going to do it. The opposing view is that held by behavioral psychologists, namely, that all of man's actions, including his thoughts, and dreams, are completely determined by various physical forces and physiological processes. The behaviorist believes that "free will" is as useless a concept for explaining behavior as divine wrath was in explaining appendicitis. This view is far too radical for most of us to stomach, yet it is one that is bound to prevail in the long run because it will lead to a tremendously complex technology that will satisfy our needs better than the technology built on the "free will" premise.

In a way, Einstein and Goedel are at least as important figures in the behavioral revolution as are, say, John B. Watson or B. F.

Skinner. Einstein pointed out that there are no absolutes in the physical world, that "truth" depends on the vantage point from which one views the universe. And Goedel proved that there is no one system of philosophy that is all encompassing. Before the twentieth century, both science and religion were philosophical systems that strived to discover the absolute structure of things. I think the greatest crisis of modern times is man's slowly growing realization that he must live with relativity not only in the physical world, but in the world of social intercourse as well. We have all been taught that there is *a* truth, a one-best-way of living and thinking and perceiving. Einstein and Goedel suggest that this absolutist approach to life is untenable. Life is fragmented, a disjointed series of episodes as bizarre and incoherent as, say, the life of the hero in the movie "Blow Up." And since I personally believe that this Antonioni motion picture is one of the most telling commentaries on modern life ever created, let me discuss it for just a moment.

You will recall that the hero of *Blow-up* is a photographer who, in the act of taking random candid shots of people in a London park, stumbles upon a corpse. Since the corpse rapidly vanishes, all the hero has left are his photographs and in developing one of them, he notices what might well be the face of the murderer lurking in the background shrubbery. He begins enlarging the photograph to larger and larger size in order to confirm his suspicions but, of course, the larger the blow-up, the more blurred the image of the face becomes. In the final shot, as he approaches the absolute resolution of the film's emulsion all he can see are almost random patterns of black dots on a white background. The face—and all other meaningful information—has completely disappeared.

When we attempt to translate knowledge about atomic physics into our everyday lives, we run into much the same sort of problem. If you don't mind humoring me for a moment, hold your right hand up in front of your face and look at it closely. Your hand appears solid, doesn't it? It surely blocks your vision of what is behind it. Now, with your left hand, press firmly on your right to see if it feels as solid as it looks. Yes, there's no doubt about it, your right hand is a very firm and real object. But of course it isn't. Your hand is made up of billions of incredibly small particles whirling around violently in an area that is for the most part empty space. From the physical point of view, your hand is very insubstantial indeed, and while it's opaque to light in the visible spectrum, it is almost completely transparent to other forms of energy, such as x-rays and neutrinos. At one level of discourse, you can appreciate the fact that the physical viewpoint towards the insubstantiality of your hand is reasonable, and yet this view doesn't help you much in reaching for a glass of beer when you're hot and thirsty. For how can one bit of almost empty space pick up and bring to your lips what is really just another bit of almost empty space?

Or, to give you perhaps another foolish-sounding illustration, suppose you are driving your car on a crowded expressway. If you think of those other automobiles hurtling along the road solely in terms of electrons whirling about a nucleus, you're not likely to survive the trip. The automobile, as you drive it along the express-way, does not seem to proceed in quantum jumps.

Have you ever tried to convince anyone with no scientific background that all matter is made up of tiny little particles? It's not an easy point to get across, as I'm sure you'll appreciate. The average man would discard atomic theory at once because it's patently obvious to his eyes that the theory is wrong. His hand *is* solid, and the hell with this academic-sounding nonsense to the contrary. All his sense organs argue for the solidity of objects, and any theoretical arguments you care to broach he will simply reject. But there is a very telling argument in favor of atomic theory that he will both appreciate and yield to. Its name is the atomic bomb. The bomb, which is a creature of that terrifying technology that was born from atomic theory, is so real and so powerful and so terrifying that it rather settles the issue once and for all. Atomic theory must be correct if it yields such dramatic products, and if our sense organs contradict the theory, well, that's relativity for you, isn't it?

When we talk about behaviorism, we should keep this illustra-tion in mind. For the usual argument against the behaviorist viewpoint is built on one theoretical, one practical and one ethical issue.

The theoretical problem that behaviorism poses is this: How can my behavior be determined when it appears so clearly to me that I have free will? The answer is, of course, that one simply cannot trust one's sense impressions to tell one what the world is really like. When one is driving on a highway, it is impractical (if not downright lethal) to pilot one's automobile according to the Heisenberg Uncertainty Principle.   Rather, one trusts one's impressions (and keeps one's eyes open). The world can be looked at from many different viewpoints, from many different levels. Any of these views may be self-sufficient, but each is a limited universe. No one view holds ultimate truth or reality; all views are possible and all are permissible. Goedel's great contribution to this problem came from his proof that all philosophical systems yield paradoxes if you press them hard enough hence, you cannot use such paradoxes as proof that the system is invalid. In short, it's quite true that you have free will and equally true that you don't, so we might as well go on to the next issue.

From a very practical point of view, one might ask a behaviorist the following question: "You claim that my behavior is entirely determined by mechanical principles, that I have no control over what I do or what I become. If this be so, why in the world are you trying so hard to convince me that you're right?" The answer to this question is as deceptively simple as is the interrogation itself: "Behavior is controlled only at a scientific or theoretical level. At the level of every-day action, free will and choice exist. To say that I feel impelled to convince you that the theory of behaviorism is valid is merely to say that I prefer the technology that stems from this theory to the technology that has come from the theory of behavior that you currently hold. In short, I'm not trying to sell you the theory except as a means of selling you the technology." Perhaps some day a more rigorous and telling argument will be put forth by some more astute logician than I am, but don't hold your breath waiting. For Western languages are based on the free-will premise, and the technology that has come from this premise is pervasive indeed. For example, as best I can see, almost all of Western art—and surely all of our culture's law—is a product of the free-will hypothesis. It is almost impossible to construct a sentence in the English language that does not have the flavor of free will about it. All our perceptions of the world (at an everyday level) are sicklied over with the pale case of individual choice, and we are as conditioned to palpitate to the word "freedom" as Pavlov's dog was conditioned to salivate when the bell sounded. Even today we could probably build a robot, made entirely of plastics and metals, that would have limited decision-making powers and be able to talk. We would, if we wished, program the robot so that it would respond to questions logically in all areas save one: If we asked it whether or not it had free will, it would always answer that it did. And the harder we pressed it with logical and empirical proof that its behavior was completely determined, the more vigorously it could be programmed to respond that its behavior was above such mundane deterministic principles. And if robots can be said to have perceptions, surely it would perceive its behavior as being "choiceful" and under its own volition. For only in those terrible moments of extreme lucidity can any of us see human behavior in scientic perspective, and surely these moments are as rare for the behaviorist as are the times that a nuclear physicist actually *perceives* the dance of the electrons in his mind's eye. Perhaps future generations will learn a different tongue and respond to different conditioned stimuli; I leave the explanatory task to them, for trying to describe the behaviorist's world to a non-behaviorist is about as easy as trying to explain quasars and pulsars to someone who still thinks the earth is flat.

The third objection typically raised against the behaviorist's position is ethical—and usually goes something like this. All right, let's agree for the moment that given enough knowledge, one person can control another person's behavior completely. Such control is obviously unethical and can be prevented only if we remain as ignorant as possible about the science of behavior. This being the case, shouldn't we shoot all behaviorists on sight and blow up their laboratories and burn their books to protect ourselves? This argument, in one guise or another, has been hurled against every scientific discipline that man has created. And the answer to it remains the same: No, science is neither ethical nor unethical. The very concept of ethics does not exist—has no meaning—in the world of science any more than the concept of an automobile has existence or meaningfulness in the world of nuclear physics. Jimmy Walker is quoted as saying that he never met a girl who had been ruined by a dirty book; in like fashion, I've never met anyone who was ruined by an equation. Ethics does, however, have tremendous importance in the world of technology, for it is the practical use that we make of knowledge that kills us or cures us.

It is at this point that the final argument usually springs forth full blown, so we might as well address ourselves to it at once. All right, you say, suppose we agree for the moment that scientific knowledge is moral—neither good nor bad, neither ethical nor unethical. Yet isn't it true that some types of knowledge are bound to lead to terrible technologies, witness the hydrogen bomb? Wouldn't it be better just to suppress that kind of scientific information right from the start, lest it be translated into action and kill us all?

You know, there's such a fatal fascination to this argument that it's difficult to make people see the fallacy behind the fascination. Scientific knowledge can be taken from the world at will by anyone who addresses himself to the problem. You might as well say to a lemming, "Look, Mister Lemming, the day may well come when you feel an irresistible urge to start swimming and find yourself plunging into the ocean and heading out to sea towards a watery grave. Under the circumstances, don't you think it would be a fine idea if we just repressed the ocean so that it doesn't lie there to tempt you to action?" Oceans, and scientific knowledge, are awfully hard to suppress, and the person who wants to do the suppressing never seems to remember one rather important fact. To make the oceans vanish, you'd have to do away with all the rivers that feed into the seas, or the rivers would just fill the ocean up again. So you end up trying to suppress not only the Atlantic and the Pacific, but the Hudson and the Mississippi and the Schuylkyll as well. And thunderstorms and rain too, of course. And even if you were successful, if somehow you managed to get rid of all the water in the world so that the poor lemmings wouldn't occasionally be tempted to drown themselves foolishly, well, what would they drink in the meantime? The sad truth is that all the lovely roses have rather vicious thorns, and if we employ biological technology to get rid of the thorns, that very technology may have built into it something else just as sharp and pointed and painful. We cannot give up technology for, in one real sense, this is what separates man from the rest of the animal kingdom. All we can do is to hope to use our technological advances in the best ways possible.

And what can behaviorism tell us about what ways of controlling human behavior are best? That is to say, can the scientific study of behavior offer any light on our ethical condition? I think it can.

To begin with, there is the power of positive reinforcement. Behaviorism is perhaps the most ethical and humane approach to the influencing of behavior because the behaviorist knows that reward is typically a stronger means of influence than is punishment. The humanists who rail against the behaviorists tend to forget this one rather crucial fact. Instead, they typically bring up two famous novels as if these pieces of fiction were telling arguments of some kind. The two books I speak of are George Orwell's *1984* and Aldous Huxley's *Brave New World*. From my own admittedly biased point of view, both novels are artistic successes but practical failures. *1984* is, at best, a rather immature prediction of what the future might be like if the new behavioral technology does not rescue us from Big Brother, although most non-behaviorists, of course, would mistakenly argue just the opposite. For in all of the past history of the human race, dictators have ruled by force, by punishment and threat of punishment. And the slaves have always revolted (eventually) and overthrown the dictator, haven't they? Well, that's just what you expect when you use punishment as your major source of behavioral control. We know from a whole host of laboratory and clinical situations that reward is a much more powerful technological tool than is punishment. The dictatorship of the future will be one in which the citizens are enslaved by love, not hate and fear and pain.

Aldous Huxley realized this fact, which is why *Brave New World* is a better book than is *1984*. But in my view, Huxley was a fink, a cop-out. He built a better mousetrap—that is, he designed a workable utopia that had a good chance to come into being—and he was so frightened by it that he had to destroy it by means of a *deus ex machina*. The first section of *Brave New World* is a masterpiece. Go back some time and read it while you're wearing a pair of Dr. Watson's Magic Behaviorism Glasses. Certainly there is inequality among the citizens of this and any other world. The alphas are smarter and larger and prettier than the epsilons, but I doubt that the alphas are all that much happier. The deltas and the epsilons work because they want to, because they are built to enjoy physical labor, and because all their needs are satisfied either by creature comforts or by that wonder drug that Huxley called soma. The thought of a political system based on love and reward was apparently so terrifying to Huxley that he had to rope in the poor confused Savage from outside to destroy the system that Huxley quite rightly feared might work only too well. No, far from promoting suppressive dictatorships, the new behavioral technology will finally give us the means to avoid "slave cultures" in the future.

The second thing that behaviorism tells us is that not all organisms are the same, that what pleasures one person may well not pleasure another. So if you want to control behavior, you have to take individual differences into account right from the start, Pigeons will work for corn as a reward, but flatworms won't. If you want to shape a human's behavior, you have to learn first what kind of reward he values most, and what his own individual response characteristics are like. Often the behaviorist is accused of wanting to turn all men into identical machines, to stamp them all from the same mold. Nonsense. The behaviorist may regard men as machines, but they're hardly identical either biologically or psychologically. Furthermore, even if men were somehow created identical, there would be no better way of programming differences into their behavior than by using the radical kind of technology I'm espousing here. If you want people to be creative, then why not use a really *effective* technology to shape or encourage their creativity? The technology, after all, doesn't tell you what kind of world you have to have; it merely gives you the tools to build the type of utopia you happen to think best. And if you happen to be hung up on utopias, or on building better worlds, let me remind you that the behavioral technology is not only the best known way of achieving your goal, it's probably the only method known for doing so.

If *1984* and *Brave New World* are poor prognostications of the future, then how does the behaviorist see it? Well, I would guess that the first changes will come in rather highly structured situations such as schools and hospitals and other institutions. For it is axiomatic that the more one controls the environment in which an organism lives, the more readily one can control the organism's behavior. The educational technology based on programmed textbooks, upon the use of computers and of classroom management techniques, is already here in prototype. In fact, our major difficulty right now is not so much in designing better teaching materials as it is in getting teachers to employ the technology already available. The usual teachers' manuals just won't do the trick, for one must change the teacher's perception of his task and convince him that behavioral technology is a viable alternative to the techniques based on the free-will hypothesis. Amazingly enough, once the teacher begins using the newer techniques, he typically finds that for the first time in his career, he is having an impact on his students—that is, for the first time in his career he's really teaching. Such an experience is usually so rewarding to the teacher that he seldom wants to go back to using the old "tried and true" methods.

In the field of mental health, the newly emerging behavioral therapies are causing great excitement. These therapies are exciting because even in their present crude and unpolished state, they offer a demonstrably more effective means of curing the "mentally ill" than we've ever had before. Now, I realize that I have made myself personally obnoxious to the bulk of the psychiatrists in this great land of ours by stating publicly what most psychiatrists prefer to mention only in private, if at all—namely, that most of the standard types of psychiatric intervention in broad use today simply don't do the job they ought to be doing. There are a great many fairly well controlled experiments in the literature suggesting that psychoanalytic and other forms of therapy probably *retard* the cure rate more than they improve it, and there are no well-controlled experiments that I know of that suggest the opposite. On the other hand, the behavioral therapists have come up with a rather large number of studies suggesting that their new techniques are far more effective than any that have been tried in the past. If we were to apply to all the mental patients in the United States even the rather crude therapeutic techniques we now have available, I would guess that at least two-thirds of the patients would be out of the hospital permanently within a year. The mental hospitals of the future are likely to be quite different from those we presently have—if, indeed, we have mental "hospitals" at all. Of course, we shall have to perform a great deal of therapy on the psychiatric nurses before this happy day comes about, but I do think we are making progress in the right direction.

But it is in the field of law that the most striking changes are likely to occur. In today's society, we seem to have criminal laws to discourage people from performing certain acts that we consider bad or anti-social and to encourage them to perform certain other acts that we consider necessary to promote society's welfare and continued existence. The primary weapon we have for enforcing these laws is threatening citizens with a variety of very unpleasant consequences if they commit a legal sin either of omission or of commission. In spite of these dire threats, the crime rate seems to soar each year, violence is on the increase, and the law enforcement agencies demand over-burgeoning budgets to deal with the situation. When you ask an American what should be done about all this, you typically get one of two responses. The liberal will say that we live in terrible times, that ours is a sick society. The conservative will say that we've molly-coddled the bastards for long enough; let's get some teeth in our laws and enforce them. Now I'd like to explore both these positions in the hope of showing that both are naive, ineffective, and hence unethical.

Take the liberal position first. Whether or not these are more frightful times than any other times we've had in the past, I don't know; but I do know that it's senseless to talk about our living in a "sick society." An ignorant society, perhaps, but hardly a sick one, for terms like "sick" and "well" are too ill-defined to be meaningfully applied to whole societies.

The liberal assumes that crime is generated by our society, that it's society's fault, as it were, not the fault of the individual who happens to commit the crime. So, you shouldn't punish the individual, you should try to change the "sick" society that spawned the crime in the first place. Now, that's a lovely but naively humanistic viewpoint that's partly right but mostly wrong. For if you ask the liberal what must be done to change society, to "cure" our culture, typically you'll get an answer that is meaningless from any practical point of view. The liberal will insist that man must learn to love his fellow man, to respect his neighbor, that we must all work hard to help each human being develop his total potential. Lovely words, but that's all they are—words. If you get down to the nitty-gritty, if you ask the liberal how he's going to implement his wonderful plan, you'll find he can offer you no more than the vaguest and most impractical of suggestions: "love other people and they'll love you," "Vote for McCarthy" or even "Get out of Viet Nam." Well, we were told at least two thousand years ago that we ought to love our neighbor as ourselves, and we've still got war and murder and prejudice. Sermons seldom accomplish anything, except to make the preacher feel good. No, good intentions are not enough, and the conservative knows this quite well.

The conservative tends to see mankind as being basically evil, born with instincts that force man to behave wickedly whenever possible. The only way to stop this innate immorality is to stamp it out wherever it raises its ugly head. Stomp on it good—that is, stomp till it hurts! The conservative believes that if we catch a criminal and beat the living hell out of him, this makes him a much better person somehow.

The conservative's typical solution to the problem of crime in the streets then is to have the streets patrolled by tanks that will blast the life out of anyone who even looks as if he's about to do something immoral. Well, in Merrie Olde England they used to chop of your hand if they caught you stealing; surely the evidence suggests that physical mutilation didn't cut down the crime rate then, nor does it now.

So, you see, the liberal blames bad behavior on the society in which the individual finds himself, while the conservative blames it on bad genes and a failure to go to church. Both viewpoints have just enough truth to them to be seductive, for man is obviously shaped in part by the society into which he's born, and punishment does at times have an influence on our behavior. But both viewpoints are also pre-scientific, and both are terribly insensitive to the actual facts of the matter. Let's see what these facts are.

The purpose of a law is to regulate human behavior, that is, to get people to want to do what we want them to do. If the law doesn't accomplish this end, it's a failure, and we might as well admit it and try something else. Laws should be goal-oriented; they must be judged by their results, or we're really being unethical and inhumane. Any time that we pass a law that more than a handful of people violate, the law is probably a bad one. Man is the only animal capable of shaping his own society, of changing his own destiny. We must use this capability to build a society in which laws become guidelines rather than threats, but guidelines with real teeth in them, guidelines so powerful that no one would want to do anything other than follow them. Probably what I've just said sounds impossible and soft-headed and perhaps even contradictory to you, but this is merely because you probably aren't as acquainted with the new behavioral technology as I am. Let me show you what I mean.

First of all, we have to reject both the liberal and the conservative tradition if we're to get anywhere. Telling people they ought to love each other, or that they ought not to be prejudiced, just isn't enough. It never has been. No, somehow we've got to learn how to *force* people to love one another, to *force* them to want to behave properly. A contradiction in terms? No, not really. For I speak of psychological force. For the most part, when we speak of using force, we think in terms of physical force, and of punishment. "Spare the rod and spoil the child," we're told. But punishment must be used as precisely and as dispassionately as a surgeon's scalpel if it is to be effective. In a way, surgery is always an admission of failure. The surgeon seldom operates on a patient unless all other means of curing the illness have been proven ineffective. Surely most physicians hope for the day when their technology is so advanced that they may abandon the scalpel forever. The behaviorist typically feels much the same way about punishment. When you're training animals, be they humans or flatworms, there are times when pain must be used as a controlling device, for in the present state of our technological development, we occasionally find ourselves in situations in which no other form of behavioral control works. But, for the most part, we try to use punishment only when we wish to remove one very specific type of behavior from an organism's response repertoire, and we try to use it very, very carefully indeed.

In contrast to this approach based on scientific data, the conservative insists that punishment be used not to control behavior, that is, to prevent crime, but rather as a kind of divine retribution to be enacted on those poor, miserable sinners who break the law. If a kid doesn't behave properly, beat the daylights out of him. That'll learn 'im. Kids ought to be bright enough to know better, eh? The conservative's viewpoint is utterly predictable to anyone who understands the relationship between frustration and aggression. It's very easy for a psychologist to devise a situation in which a laboratory animal is intensely frustrated. Under the right conditions, the frustrated beast quite predictably turns on and attacks any scapegoat that happens to be handy. When humans are frustrated, they typically become aggressive. Thus, when some poor demented fool assassinates a beloved leader, the crowd usually turns on the assassin and tears him limb from limb. The crowd is frustrated by the assassination—they don't understand why the man was killed, and they are powerless to restore life to him. The crowd's immediate response to this frustrating situation is aggressive revenge. When lawmakers don't understand some aspect of the world around them, and when they are frustrated by something the people do, or the President does, or the Supreme Court rules, the lawmakers typically respond by passing a highly punitive and aggressive law. Yet these are the very situations in which punishment has little or no effect on the behavior *of the very people* the lawmakers want to influence or control. And so bad laws get written simply because they make the lawmakers feel good.

Parenthetically, the effectiveness of positive reinforcement as a tool for shaping human behavior seems intimately linked to what I call "the illusion of free will." When we use punishment to shape an individual's responses, what we really communicate to the person is the following message: "Either you do what I tell you to do, or I'll hurt you badly." The subject of such a threat typically does not see himself as having much choice or free will in this kind of situation, and he usually resents being (quite literally) forced to do your bidding. When you offer him positive reinforcement for changing his behavior, though, he often does not perceive the threat to his freedom as being nearly as great. For anyone can resist pleasure, can't he?   Denying yourself pleasure just isn't in the same ballpark as exposing yourself to severe punishment. And yet, the experimental data suggest that positive reinforcement is really much more effectively coercive than is negative reinforcement.

For instance, suppose you were called upon to handle the following real-life problem. A fourteen-year-old delinquent girl gets into trouble with her family, her teachers, and even with the police, because of the foul language she uses continuously. Her speech is so loaded with four-letter Anglo-Saxon words that even a sailor might pick up a few choice phrases from her if he listened for a minute or two. How would you go about getting the girl to "talk decent"? The typical approach would be to give her an electric shock, or some other punishment, each time she used a dirty word; but, as Richard B. Stuart (a colleague of mine in Michigan) points out, punishment simply won't work with this girl. In fact, punishment may actually reward her for it tells her that "she's getting to you," and she rather likes upsetting her elders. At best you could temporarily suppress the cursing by punishing it, but you would also be building resentment towards you at the same time and eventually the girl simply throws a temper tantrum. Instead, Professor Stuart suggests, you should give the girl a tangible reward for not swearing. At the beginning, you might give her a nickel if she can talk for three minutes without uttering a dirty word. When she has earned a few nickles, then you increase the time period to six minutes and the reward to a dime. When she has shown herself capable of earning the dime, you increase the time period again and up the reward appropriately. Each time she earns the reward, of course, you praise her lavishly, and when the girl is able to go an hour or more without using offensive language, you begin to reward her only intermittently with money but continue the praise. Eventually, the girl will be able to control her speech in public just like anyone else, even when no one is around to give her money, and she will be reasonably happy to do so. In point of fact, the girl rather enjoys "the nickel game" and asks to play it every chance she gets. Whatever minor qualms the girl has about your "bribing" her to behave are readily soothed by the purchasing power of the money she earns. She thinks she has a choice, that she can go back to her old way of behaving if she really wishes to, but this is an illusion on her part for she is, in fact, under much stronger social control then if punishment were used. Thus it is important that whatever techniques we employ to modify behavior have the illusion of free will and choice built into them, but it would surely be inefficient (hence, unethical) if these techniques offered people more than a mere illusion.

My own belief is that the day has come when we can use such radical techniques as the complete social and physical isolation, as well as drugs, hypnosis and the astute manipulation of reward and punishment, to gain almost complete control over an individual's behavior. It should be possible then to achieve a very rapid and highly effective type of "positive brainwashing" that would allow us to change dramatically a person's behavior, and his personality. No one knows at the moment what the limits of such change would be; my own guess is that the limits would be as few and as far between as those that apply to nuclear energy. I foresee the day when we could take the worst criminal you can imagine and convert him into a decent, respectable citizen in a matter of a few months—or perhaps even less time than that. The danger is, of course, that we could also do the opposite—we could use these techniques to change any decent, respectable citizen into any kind of criminal you'd care to mention.

My proposal then is this: We must begin by setting our legal house in order. We must draft a new set of laws that will be as consonant as possible with all the data on human behavior that scientists have gathered over the years. These new laws would be written with the techniques of behavioral control firmly in mind; that is, we would try to regulate human conduct as much as possible by offering citizens rewards for "good" behavior rather than by threatening them with punishment if they broke a law. We would want to reshape our society drastically, so that all of us would be trained from birth to want to do what society wanted us to do. I firmly believe that we have the techniques presently available to undertake that kind of cradle-to-grave education, and that only by the use of these techniques can we ever hope to maximize human potentiality. Of course, as I suggested earlier, we cannot give up punishment entirely, but we can use it sparingly, intelligently, as a means of shaping people's behavior rather than as a means of releasing our own aggressive tendencies.

As I see it, in the case of misdemeanors or minor offenses, we would administer as brief and as painless a punishment as would achieve the end of stamping out the anti-social behavior this particular criminal displayed. Coupled with this would be an attempt to restructure society so that we could give potential or actual criminals such a wide array of socially accepted behaviors to choose from, and reward these behaviors so vigorously, that few if any of them would be tempted to stray from the straight and narrow. For major offenses or felonies, we'd take a different tack. We'd assume that a felony was clear evidence that the criminal involved has somehow acquired a full-blown social neurosis and needed to be "cured," not punished. Locking the criminal up in a prison is inhumane, costly, and wasteful of human potential, for it is the anti-social *behavior* we wish to remove, not the person himself. So we'd send a criminal convicted of a major crime to a rehabilitation center where he'd undergo "positive brainwashing" until we were quite sure that we had turned him into a law-abiding citizen who would never again commit an anti-social act. At the beginning, at least, we'd probably have to restructure the criminal's personality entirely. Many of you will rebel against such a notion, for you still cling to the old-fashioned belief that each of us builds up his personality logically and by free will, a notion that is as patently incorrect by scientific standards as the belief that the world is flat. The legal and moral issues raised by such procedures are frighteningly complex, of course, but surely we know by now that there are no simple solutions to the difficult problems that face society today.

What I'm suggesting, of course, is that no one owns his own personality. Your ego, or individuality, was forced on you by your genetic constitution and by the society into which you were born. You had no say about what kind of personality you acquired, and there's no reason to believe you should have the right to refuse to acquire a new personality if your old one is anti-social. Most of us realize, however dimly, that we do not own our own bodies. The government has the power to incarcerate us, to punish us, to forcibly cure us if we have a communicable disease, to send us off to get killed in battle, and even to execute us. When we also realize that our personalities are no more than behavioral extensions of our bodily processes, it should become clear to us that we have, at best, squatters' rights to our egos and that society has just as much right to control our minds as it has to control our bodies. If we insist that our spirits are free, we are merely deluding ourselves with pre-scientific thought, and no matter how comforting that thought may be, it's patently illogical and, in the long run, it's probably unethical too. For example, if you insist that society keep its hands off your thought processes, that it leave you with complete free will over your own behavior, then you are insisting that you ought to be able to commit a crime whenever you jolly well feel like it. I don't believe that the Constitution of the United States gives you the inalienable *right* to commit a crime if you want to; therefore, the Constitution does not (to my way of thinking) guarantee you the right to maintain inviolable the personality it forced on you in the first place—if and when that personality manifests strongly anti-social behavior.

In a recent article, David Bakan points out an interesting parallel between the rise of behaviorism in the United States and the rise of the Protestant Ethic. According to Professor Bakan, most of the great behaviorists have been white, Anglo-Saxon Protestants from rural areas or small towns in southern or mid-western states. Since I am a white, Anglo-Saxon born of Protestant parents in a small midwestern town, perhaps it is no wonder that I became a behaviorist who believes that man can save himself through good works. Modern day physics had its birth in Western Europe rather than in the Far East because the concepts of the physical sciences were acceptable to European philosophy but were probably unthinkable to a Buddhist. Does this fact invalidate nuclear physics? My own personal WASPish belief is that ethics is merely a guideline for acceptable behavior between or among two or more individuals. Thus ethical behavior is always goal-oriented and pragmatic. Indeed, I say of ethics what others have said of politics—namely, that it is the art of the possible. In any system of ethics, whatever behaviors achieve the goal best and fastest are seen as being good and righteous; whatever behaviors lead away from the stated goal or achieve the goal only very slowly are typically seen as being bad and unethical. Behaviorism cannot tell us what goals should be, but it can tell us how best to achieve our goals once they are decided upon and, by encouraging us to apply scientific methodology to the solution of human problems, might even help us evaluate the goals themselves.

I do not know what humanity's goals should be, but I do think a couple of things are obvious. In the first place, one of our primary aims must be the survival of the species and, to a lesser extent, the survival of some kind of culture of society. I would guess that both the species and the society are in a continual process of evolution and adaptation and that these processes of gradual change are worthwhile. Any technological advances that increase the probability of survival of the species and/or the culture, therefore, are likely to be adopted by humanity. Since the new behavioral technology appears to offer us a greater chance of survival than does the old, it will surely become a part of our world in the near future. And since, in my opinion, ethical systems stem from practical considerations, I predict that what this generation calls "radical behaviorism" and therefore rejects, future generations will accept as being much more "ethical" than the rather fuzzy-minded, humanistic code of ethics that dominates our thinking today.

CHAPTER THREE ✳ BEHAVIOR ✳ *The Stick and the Carrot*
by V. G. DETHIER

According to an old fable the most successful way to get a donkey to do one's bidding is to hand a carrot out in front of him and wield a stick behind. Despite the unscientific nature of this fable, or perhaps because of it, the fable may contain more wisdom and pragmatism than is provided by psychology. Since there is a little bit of the donkey and more than a little of the jackass in each of us, the relevance of the story to the thesis presented by Dr. McConnell will become apparent. Both fable and thesis deal with the control of behavior, with positive and negative reinforcement, and with free will (although in the quadruped donkey it would have to go by a different name).

Dr. McConnell has presented us with a dream that we all wistfully wish were possible of achievement. He has proposed means of achieving it. To wish for it and to achieve it are, however, two quite different accomplishments. I have no quarrel with his goal, but I would like to add some cautionary notes about the means. He would construct a better world from the eminence of two premises. The first is that theoretically, at least, we can completely control a person's behavior. The second is that reward is a typically stronger influence than punishment. How strong is the evidence in support of these two beliefs? What is the probability of their being correct?

The fable about the donkey continues. In one instance when nothing would move the beast the driver built a fire under it, whereupon the donkey stepped forward just far enough to remove his belly from the heat of the fire and place the wagon over it. Since nothing would induce the donkey to progress farther, the wagon burned to a cinder. In another instance a donkey that carried loads of salt for his master developed the clever habit of lying down in a stream that had to be crossed on the way to market. Naturally, the salt dissolved thus lightening the load. The solvency of the salt resulted in the insolvency of the master who thereupon decided to teach the donkey a lesson. He loaded him with sponges on the next trip. This time when the beast lay down his load increased in weight, as a matter of fact, to such an extent that he was unable to rise, so drowned.

Now I am dealing with more than analogy here and if I choose to derive conclusions from certain behavioral techniques applied to a domesticated odd-toed ungulate, are these conclusions any less vulnerable than those derived from techniques applied to an inbred domesticated rodent? To what extent do they have relevance to man? Most of what psychologists know about behavior is derived from studies of the white rat. About twenty years ago Beach defined psychology as "the study of learning in the white rat and the college sophomore." As recently as eight years ago the Brelands pointed out that Beach's admonition to extend investigation beyond the white rat had been largely ignored by psychologists despite a continuing realization that the white rat may not be able to reveal everything that is to be known about behavior. Although recent years have witnessed a growing interest in the behavior of a wider spectrum of species, the bedrock upon which most of our understanding of learning behavior now stands is a rubble of data derived from studies of learning in the rat. I think that it is premature, and even presumptuous, to try to apply this psychology of learning to the manipulation of human behavior.

Even if data obtained from studies of a highly inbred, domesticated rodent were extrapolatable to man, how much reliance can be placed upon a conditioning technique? The Brelands, who have conditioned over six thousand animals comprising thirty-eight species, began to encounter a persistent pattern of failures. As they remarked, "These egregious failures came as a rather considerable shock to us, for there was nothing in our background in behaviorism to prepare us for such gross inabilities to predict and control the behavior of animals with which we had been working for years." Here the fables about the donkey take on an aura of reality. The Brelands conclude, reluctantly, that the behavior of a species "cannot be adequately understood, predicted, or controlled without knowledge of its instinctive patterns, evolutionary history, and ecological niche." To which, for man, one must add cultural and social history.

In short, our understanding of behavior in general, and the behavior of man in particular, is so fragmentary, so specialized, and so strongly influenced by restricted theory derived from highly esoteric studies, that I for one hesitate to predict that the day will come when we can control man's behavior. I am employing the word control to mean manipulate according to plan. Not only do we know so little about behavior that we cannot yet control it, we know so little that we cannot say with conviction that it is controllable. To assert the contrary is to make an act of faith no less wondrous than the act of faith that belief in God requires. Furthermore, even if one were able to control the behavior of an individual, it does not follow that one could control groups of animals. In animal experimentation man is outside the experiment; in controlling the behavior of groups of humans in the brave new world, man is inevitably part of the experiment. We cannot control ourselves, let alone control others. We lack not only self-control, but the knowledge, the wisdom, the altruism, the compassion. The response of a number of psychologists at this year's annual meeting of the APA to questions by reporters from *Science* is thought-provoking. They said that the relevance of psychology to present-day problems is highly questionable.

Dr. McConnell's second premise is as debatable as his first. He has told us that "reward is a typically stronger influence than punishment." In this connection one would do well to read the 1963 Presidential Address to the Eastern Psychological Association. In it Solomon sets out to decry some unscientific legends about punishment. One of the most persuasive advocates of reward as opposed to punishment for controlling behavior was Kinner who in *Walden Two* would use no punishment because it produces poor behavioral control. Solomon, while admiring the humanitarian and kindly dispositions in such writings, points out that the scientific basis for such conclusions is shabby. I would rest my case here by quoting a concluding paragraph from Solomon's address.

"Our laboratory knowledge of the effects of punishment on instrumental and emotional behavior is still rudimentary—much too rudimentary to make an intelligent choice among conflicting ideas about it. The polarized doctrines are probably inadequate and in error. The popularized Skinneran position concerning the inadequacy of punishment in suppressing *instrumental* behavior is, if correct at all, only conditionally correct. The Freudian position pointing to pain or trauma as an agent for the pervasive and long-lasting distortion of *affective* behavior is equally questionable, and only conditionally correct."

No one can be certain at this moment that the carrot is more effective than the stick. It requires both.

None of the foregoing denies the power of reward. It does raise the puzzling question of what constitutes reward? Reward is pleasure; however, my pleasure may be anathema to you. My most exquisite pleasure, beyond which there is no greater reward, may be the very anti-social behavior that you are trying to teach out of me. A more powerful argument, however, concerning the identity of reward hinges on goals and aims. What I consider a reward depends in no small measure on what I want out of life. If all I desire are wine, women, and song, and the four freedoms, he who would shape my behavior by reward has, theoretically at least, the means at hand. I may, however, see my role on earth in a different light. Dr. McConnell wisely steers clear of aims and goals. He does hazard a guess that survival of the species and of a society are the important aims of man. All of this may be true of men who do not believe in a hereafter but, like it or not, a sizable fraction of the world population does believe in a hereafter or in God or gods. Jews, Christians, and Mohammedans believe in God. Hindus believe there is another life. Even those who practice Shinto believe in another life as evidenced by ancestor worship. Now it does not matter a whit whether or not the behaviorist believes in these things. A large number of people in the world, as a roll call of the adherents of the various religions would attest, do believe in these things. For them, then, the supreme rewards may lie beyond the reach of the psychologist. The aims of man could have to be changed in order to be able to reward him as he wished. One of the difficulties in effecting these changes lies in the fact that society is composed of many generations. The controlling of behavior would have to begin somewhere. It is all right to begin at the cradle, but how does one remove the influences of the parent who has not yet been brain-washed—by taking away the children?

Under a democracy it cannot be done. Even under dictatorships it has failed.

So it is that I am regretfully in disagreement with Dr. McConnell's belief. The world is indeed in a mess. I don't believe we are in a Behavioral Revolution. Man is perceiving himself differently than before; he is in a perceptual revolution, but is behaving much as always. At the recent annual meeting of the American Psychological Association many psychologists interviewed by *Science* felt that psychology could do little to alleviate the distress of society. Whether this is due to the impotence of psychology or to its irrelevance to the troubles that smite us on all sides, I would not presume to say. My own conclusion is that we know too little about animal behavior, less about human behavior.

I feel somewhat like Alice when she said, "There's no use trying, one can't believe impossible things" despite the White Queen's reproachful "I dare say you haven't had much practice." I feel that dreams are impossible when they derive from incomplete knowledge or error. I agree that the world would lie unpleasantly stagnant if someone did not dream the impossible dream, but man has not yet done his best in striving to achieve the possible.

CHAPTER THREE ✳ BEHAVIOR ✳ *Whose Will Will Work?* by
MARTIN W. SCHEIN

It's hard to take issue with much of what Dr. McConnell has presented. However, some items that he threw in for shock value are somewhat less than shocking. For example, he started out by saying that we now can (theoretically at least) "gain such complete control over a person's behavior that we should be able to change him from whatever he is to whatever we think he ought to be." Despite our very imperfect bumbling, empirically established methods, controlling others' behaviors has been one of our prime preoccupations since at least the dawn of recorded history. I might also say that, our very poor technology notwithstanding, we've enjoyed a modicum of success: few of us engage in wholesale or even retail rape, theft, or murder, despite the sometimes severe provocations offered by miniskirts, store windows, and mothers-in-law.

Nevertheless, Dr. McConnell makes a good point, and it should not be lost in clever verbiage. The point is that the state of the art in behavioral work is so advanced these days, relative to what it was a few short years ago, that we can now mold someone's behavior far more efficiently and far more easily than could have been done before. The difference is quantitative, not qualitative. I agree that a technology based on what we now know is only in its infancy, and our children or grandchildren will enjoy the fruits of the trees that will sprout from the seeds now being dispersed. I am referring, of course, to a judicial system based primarily on reward rather than punishment, to an education system where learning is far more important than teaching, and to a social system aiming at life with dignity rather than mere survival.

Dr. McConnell's point on behavior mechanisms over free will is well taken, but I believe he should have been far more shocking here than he chose to be. As the basic mechanisms of behavior become more fully undertstood, so do they become more amenable to control, *not* through the devices of reward and punishment but through the buttons that manipulate gene complexes or electrochemical signals. The science fiction fantasy that depicts a madman (or even a sane man) sitting in front of a console and controlling, by means of appropriate buttons, the behavior of masses of people outside is all too frighteningly nearing nonfiction.

The technique is alarmingly simple: non painful, non-debilitating electrical or chemical stimulation applied to appropriate areas of the brain have gotten angry bulls to instantaneously stop mad charges; otherwise occupied roosters to suddenly crow; nondescript chickens to attack, retreat or assume sleep posutres; and normal female turkeys to display male courtship postures to appropriate models. I have done some of this work myself, and I watched one great researcher manipulate via telestimulation techniques the social behaviors of several chickens in a flock outside. The science fiction picture of the madman at the console is not too distant. Humans are not proscribed from such studies: intracranial stimulation applied to specific areas of the awake human have alternately produced feelings of euphoria and rage, and have evoked old forgotten memory traces.

So, if you'll pardon the pun, here is something of real shock value. What would happen if the newfound ability to readily control behavior by simple electrochemical means were turned to devious channels; if the technology became a powerful weapon in the hands of an ever-present ruthless tyrant? No longer does he have to destroy the enemy that impedes his progress: he merely has to "tune them in" to his purposes. No messy corpses to deal with, no blood and, more importantly, he can still get a lifetime of useful work out of them. It wouldn't be so bad if the ruthless tyrant were you or me, but it would be a calamity if the buttons fell into the hands of the other party or our worst enemies.

So what are we to do? We can, *for once,* look at a situation rationally before it becomes a major problem, since responsibility to society implies also responsibility on the part of society. Something is coming along that will have a profound effect on human behavior, and society had best consider right now the technologies that will flow from this work. The potential benefits are great, as are the potential threats. The answer is *not* to stop the research at this point, for as Dr. McConnell has pointed out we simply cannot stop (nor even impede for long) the pace nor direction of scientific endeavor. Nuclear fission is neither inherently good nor bad: it is the use to which it is put that is desirable or undesirable. In the same way, our understanding of behavioral mechanisms is neither good nor bad: it is the way we as a society utilize the knowledge that is beneficial or reprehensible.

The challenge is to society to *act* with forethought and deliberation, rather than *react* with panic and frustration.

CHAPTER THREE ✳ BEHAVIOR ✳ *No Free Will?* by C. RICHARD SNYDER

I should like to begin by relating to reactions of two psychologists to Dr. McConnell's ideas. One, relatively young, calls himself a "humanist" and says that Dr. McConnell hasn't gone nearly far enough, that his techniques could be extended far beyond the realms suggested by the paper. The other man, who refuses to wear a label of any sort, and who is the senior of the two, believes that Dr. McConnell and all behaviorists, for that matter, should exercise more humility and more restraint. The ideas in the paper, while they may be valid under controlled conditions, may not be free of harmful side effects when exercised in the field of the outside world of human existence. In other words, we biologists would say that what works *in vitro* doesn't necessarily always work *in vivo*.

The remarks that follow are personal—they are the reactions of a teaching biologist who is not a psychologist.

Dr. McConnell envisions a society in which every member will be an ethical being. Anti-social behavior can be changed to actions which result in the "best for all" because man has no free will, and so the changes necessary can be accomplished easily and quickly by the application of the "behaviorist technology."

The behaviorists, he contends, are the only practitioners of this technology, and it is into their hands that society must place itself if our species and its culture are to survive.

This technique could change criminals into law-abiding citizens in a few months. It would change every man who needed to be changed, not turing all out of the same mold, but permitting complete expression of creativity by each individual. It would "force" people to become ethical, by giving them the illusion of "free will."

Specific methods would include social and physical isolation, drugs, hypnosis, and the "astute manipulation of reward and punishment." Some offenders would receive light punishment for misdemeanors, others would be bribed to abandon their anti-social actions, while convicted felons would be given intensive treatment at rehabilitation centers—a sort of "positive brainwashing."

Several of these ideas require careful scrutiny. Perhaps the most controversial proposition is that man is without "free will," except at the level of everyday action. Dr. McConnell admits that this viewpoint will be difficult for a non-behaviorist to swallow. For me, it is impossible. I'd like to think that I've had something to say, by the expression of my free will, over what has befallen me, for good or evil, over the years.

If man has no long-range free will, and the "bootstraps" philosophy be completely nullified, then the entire Community College movement is founded on a non existent belief. We hold that there are some students who can, in spite of past failures, become useful and productive members of an academic communi-ty, and that they should be given the chance to do this. Many of them, by means that transcend genetic and earlier environmental influences, meet with success in the institution in which I teach and go on to four-year colleges, and even into graduate and professional institutions. In discussing such cases, we often use terms like "maturity," "effort," and "application," all of which presuppose the existence of free will.

I cannot accept the proposition that our future is secure only within the framework of the behaviorist technology and I would even suggest that much of our recent national violence stems from the implementation of these very ideas. The leaders of some of the most destructive campus activities were reared in permissive homes. While it is true that the behaviorist technology, in Dr. McConnell's interpretation, does not include complete permissiveness, the two philosophies have much in common. For example, if man has no free will, he can't be held accountable for his actions, can he? If this is so, what right does society have to punish him?

One final troubling point. The behaviorists do not, they say, set goals; their technology merely sets guidelines for reaching the goals set by others. It would seem that a statement made at the annual convention of the American Psychological Association in Washington at the end of August would contradict this claim completely. The psychologists, according to the news release of August 31, 1969, want a piece of the action with respect to domestic and foreign political affairs. They believe that they should be consulted by the politicians to a much larger extent than has been the case. They claim to possess knowledge about human behavior that could have prevented at least half of the major wars of recent times. In other words, at least some behaviorists want to help set goals for human actions.

We have institutions which can provide the dual functions of setting goals and providing pathways to these goals. True, the church, the home, and the school have not realized their potentials in this respect. This is particularly true of the first two of these.

The church, once a bastion of complete conservatism, has swung over to the other side, and now there are services in which the music is supplied by three guitars and a drum, the liturgy replaced by the reading of the works of one of the poets of the "now" generation, and the sermon supplanted by a "conversational encounter" between the pastor and a youth, or between several youths—always youths, of course. These things are being done on a very narrow scale, it is true, and their real worth, in terms of achieving ethical standards of conduct, has yet to be demonstrated.

Many homes have given up the battle completely. The two generations in the home are equally sure that there is no point of contact, that the "generation gap" has become impossible to bridge.

That more or less leaves things up to the schools. Dr. Carl P. Rogers summarizes the role of the schools when he writes: "The challenge for education is unreal only if we are looking a year or two ahead. From the long view I know of no problem holding greater potentiality of growth and of destruction than the question of how to live with the increasing power the behavioral sciences will place in our hands and the hands of our children." As my one psychologist colleague puts it: "Here's a Pandora's box. If you must peek, be prepared to be responsible for the results."

CHAPTER THREE ✻ *Discussion*

QUESTION—*How will long periods of travel in space affect man psychologically?*

DR. McCONNELL: I don't really know. I think this is what psychologists would say is an empirical question and the answer will come from experimentation. We'll have to try it and see. I think that what we will probably do is see what the effects are and then try to find ways to overcome whatever defects are discovered. I don't see greater differences in behavior of people in isolation, as in a space capsule, than in any other situation with a limit on incoming sensory stimulation.

QUESTION—*If man is not free, is he responsible?*

MR. SNYDER: The statement I made went something like this: If man has no free will, he can't be held accountable for his actions can he? If this is so, what right does society have to punish him? That requires an iffy answer. In other words, I said originally that I believe that man is free. Therefore, I believe that he should be subjected to the limitations that society imposes upon his right to disturb the equanimity of others. I'd also like to make a comment about what Dr. McConnell said in his discussion that nuclear devices were prepared by physicists but their destructive utilization was decided by others. Others means politicians in this case. And then he goes on to say the use of these behaviorist techniques would be in the hands of society. All I can tell you is that if he equates society with the politicians, then God help us all.

DR. McCONNELL: The question of responsibility is a difficult one, as I tried to imply. There are two answers: yes, man is responsible obviously at one level—that is the everyday level; but at the theoretical level he is not. One can see the same thing happening in biological terms. If I am exposed to someone who has had smallpox and I have been vaccinated and I catch smallpox, am I responsible for having smallpox? Is it my responsibility? The answer to that is yes and no, if you wish, but it certainly is right for society to isolate me, or to cure me, or to force me to get well. I think that what society attempts to do if I caught a bad idea (which is evidenced in bad behavior) is not talk about responsibility, or lack of responsibility, but try to get rid of the bad behavior. We don't punish people deliberately simply because they were stupid enough to catch smallpox. It is not a matter of punishing them, it is a matter of getting rid of the smallpox. The same thing is true with bad behavior: one tries to get rid of the bad behavior and not get rid of the person; if you want to call that responsibility, fine. I am responsible because I caught it. Now let's get rid of it.

QUESTION—*How can one find suitable rewards for all the various contingencies in different kinds of behavior one might wish to eliminate?*

DR. McCONNELL: I wish I knew the answer to that. What we do when we deal with real-life clients is to ask them what they want. It is a very simple-minded way of going about things. You can't always trust their answers, but it is a good first approximation. Simply begin by saying, what do you want, and it sometimes turns out that what they say they want is, in fact, incompatible with what they really want. Then you have to deal with the matter in a more indirect way. I can't say entirely what motivates people—

there are obvious biological things that we want. When faced with a severe behavior problem it may be useful to see what has been learned in dealing with juvenile delinquents. Sometimes when you get a kid who is so messed up and who comes from a family that is so messed up that you cannot deal with the family and the kid at the same time, the only thing you can do is pull the child out, isolate him, and put him in a structured environment such as a detention home or housing facility where you can control his biological needs and where he has to work in order to get food and a place to live. I speak here from my experience with a detention home we have like this in Ann Arbor which works beautifully. Most children tend to thrive in this kind of environment and you can make some rather major changes in their personalities and in their ways of behaving which last as long as they are in that environment. Unfortunately, when they return to their crazy home, they usually return to their crazy behavior. That is a bit of a problem. Then you have to find some way of controlling the parents' behavior, and that can be much more difficult because you don't have quite the legal control that you do over the kids. We are always looking for ideas about what rewards might be used because sometimes it is very difficult to find effective ones.

QUESTION—*Please comment on the increasing use of drugs by high school and college students and on better ways of dealing with the problem than currently are being employed.*

DR. McCONNELL: I think that the present Operation Intercept is a farce and I don't think it is going to achieve anything demonstrable. The use of punishment as it is currently employed is, I think, more political ploy than anything else. Maybe a political ploy has use for some people but I don't think it will be useful in changing drug use. I will also admit quite frankly that this is one area in which the behaviorist has real difficulties, probably because many of them are not as biological as I would say Dr. Dethier would want us to be. It is not too difficult to deal with high school kids who are on pot, as it doesn't have quite the physiological effects that some of the other drugs do, so we can usually find adequate rewards to substitute for the smoking of marijuana. That is not so much the problem. The real difficulty is with methedrine, with speed, and the amphetamines. So far as I know, Dr. Stuart, who has done work in this area, has been completely unable to take kids who want to get off these drugs and find any way of helping them to get off. Most drug taking, like smoking pot, occurs with certain types of stimulation, certain environmental conditions, which facilitates or promotes smoking pot or taking LSD or something like that. This social aspect doesn't seem to be true of the amphetamines. We have not yet been able to discover those conditions, those stimuli, which trigger off the biological effects, and until we do we are going to be in trouble. We are doing work on this now, but I see no easy solution to the problem at this time.

QUESTION—*Please comment about parapsychology, such as that conducted at J. B. Rhine's Laboratory at Duke University.*

DR. DETHIER: Since I am not a psychologist I can say things that a respectable psychologist can't say, but as far as parapsychology is concerned, I can't speak with any authority at all.

DR. SCHIEN: I am not a psychologist either, so I can speak freely and I can quote all of my psychologist friends. I have yet to find any one of them who holds this work in high regard. I've even talked to statisticians about it and there is, perhaps some statistical hanky-panky. But I have never found any one who had held this work in high regard, including people who are at Duke.

QUESTION—*What is parapsychology?*

DR. DETHIER: Parapsychology is concerned with extra-sensory perceptions and the ability of individuals to predict things which presumably might happen in the future.

QUESTION—*When the particular subject is fully rewarded, what further reward can be used to modify behavior?*

DR. McCONNELL: Dr. Ivar Lovaas of UCLA is working with autistic children and doing a very remarkable job using extrinsic rewards, like ice cream, to get children to behave in better ways than anyone has been able to do up until this time. In talking to Dr. Lovaas about it, he pointed out one very interesting thing. He began working with children, using ice cream and candy as rewards, and he said that ice cream works for about six weeks and then it no longer is good as a re-enforcer. But, he said, the kids never satiate on meat and potatoes. And I think this goes back to what Dr. Dethier was describing earlier—it is necessary to take the biological status of the organism into account. One of the reasons one tries, as I suggested earlier, to shift from extrinsic rewards like food or nickels, to social rewards, is that it is far more difficult to get satiated on social rewards than it is on biological rewards. I can only give you my own personal insight into this, but I almost never tire of praise.

QUESTION—*Are any patterns of anti-social behavior inherited?*

DR. McCONNELL: I'm really not typical of the behaviorist because Dr. B. F. Skinner, who I guess is the prototype, doesn't believe that physiology exists and I do. I think the genes are there but there is nothing in the behavioral literature to suggest that we have any strong evidence that types of anti-social behavior are coded in the DNA molecule and that one inherits them. I do know that a fair amount of evidence coming out of the Boston area at the moment suggests that various types of strongly anti-social behavior, aggressive behavior and hostility are highly correlated with hyperactivity in certain parts of the brain, particularly in hippocampus, or medulla. In short, what I am trying to say is yes, there are biological underpinnings. But I don't think the data or the XYY chromosome correlation with criminology is very good. I don't believe it is going to stand up, but I do believe that there are some types of behavior patterns, which we would consider anti-social, that may well be correlated in some way with some kind of chromosome damage. The difficulty is at a physiological level and, of course, this places a limit on what you can do behavioristically. If you'll buy a little substitute, put feathers on your human and call him a chicken. Gamecocks are perfect examples of breeding for aggressiveness; they are rather vicious animals. I think with a couple of minutes thought we could probably come up with a great many other examples where you can tie behavioral complexes down to the genetics of the animal breed. In fact, you can take an ordinary chicken, or an ordinary whatever species you want to deal with, and start selecting for whatever behavior complex you want and usually within a few generations you will begin to get it.

QUESTION—*What is the evidence to indicate that reward is more effective than punishment?*

DR. McCONNELL: There is a whole range of studies that Dr. Dethier referred to, at the animal level. One way or another, if you want to go back and look at them, I suppose the best thing to do is to start with some of Skinner's work and see what he has to say. I don't want to be put in the position of saying that positive reinforcement is the only thing one should use. As I think I made clear, I want to apply the stick every once in a while, too, but I want to apply it to a specific behind and for a specific condition. The sort of thing we find in working with humans is that typically (and these are not controlled studies necessarily) when you punish people they resent it, and when you reward them they like it. They come back for more. The behaviorist who is trying to shape behavior likes that, because it makes his job a lot easier. A little girl we gave nickels to, as I said, likes the game; she wants to keep playing it. When we rewarded her negatively, that is, when we gave her punishment for doing bad things, she ran away. She went all the way to California. That taught us a lesson. You can't, in Michigan, control the behavior of a little girl when she's in California. So I think these are the sort of data *in vivo,* rather than *in vitro,* that lead most of us to believe that whenever we can use reward it is a better thing to use.

QUESTION—*If the white rat is not the best possible animal to use to analogize the man, is there a better one?*

DR. DETHIER: I think there is a better animal and the better animal is man. Man in the generic sense. I happen to work with flies; Dr. McConnell has worked with planarians and is now working with delinquent girls. I wasn't trying to indicate that I don't think the work being done wasn't good work. I was trying to make the point that in perusing the literature of psychology upon which many of our ideas of behavior are based, one finds a preponderance of the work has been done with certain specific animals. I merely raise the question as to how sound the conclusions are, if you wish to apply them to man, when they are based on a very small number of selected animals especially in view of the fact that there are over a million species of animals in the world. It is not a question of which animal species is best. I think we should do more with people within the limits of what is now legally permissible.

QUESTION—*How is behavioral therapy used in the treatment of mental illness?*

DR. McCONNELL: I wrote an article, "Psychoanalysis Must Go," that appeared in *Esquire* a year ago which gives some of the history of how behavior therapy came into being and where it may or may not go. The general idea is that most forms of what we call mental illness are due to bad learning. It is not that someone is mentally ill, but rather that he has acquired bad behavior or sick behavior or insane behavior over the years. It is pretty hard to cure the mind of mental illness; nobody seems to be able to do it. It is fairly easy, under some conditions, to cure certain types of behavioral syndromes by simply deciding what the patient ought to do to show well behavior rather than sick behavior, and by rewarding any sign of what we would call well behavior and punishing sick behavior if necessary; it is a long and very complicated process that involves a fair amount of technology. A final comment is that it is dangerous to generalize from flatworms to delinquent girls. I discovered this because of their biological differences. I have never found a delinquent girl who was at all flat.

CHAPTER FOUR ✳ GENETICS

CHAPTER FOUR ✳ GENETICS ✳ *Genetics and Genetic Manipulation* by BRUCE WALLACE

The participants in this Symposium are attempting to break a tradition: namely, that science teachers should limit their classroom comments solely to observable and confirmable facts. Events move rapidly in today's world. In the past, the evaluation of a new scientific procedure would be carried out in a leisurely fashion; the exposure of persons to new industrial products or to new medical practices proceeded one or two persons at a time so that faulty practices could be abandoned before they caused serious harm to entire communities. Today this time for evaluation and reflection, time for changing one's mind, has all but disappeared. Enormous sums are spent on medical research; huge firms stand by to mass-produce new medicines and vaccines; thousands of persons may be involved in preliminary trials alone. The elapsed time from discovery of a vaccine to its general use throughout the country may be remarkably short. Decisions, once implemented, frequently have irreversible consequences. It is only after the fact, for example, that we learn that a widely used polio vaccine contained, in addition to the polio virus, a simian virus of unknown effect on human beings; at best, we can now only wait and see what the outcome of this oversight will be.

The shortened time between discovering and doing, the ease with which large segments of the population can be swept up and involved, and the irreversibility of many actions make it imperative that, simultaneous with the discovery of new techniques, their social implications can be discussed in science courses. Value concepts lie beyond scientific proof and so the science teacher may not be able to prove to everyone's satisfaction that this or that consequence is "good" or "bad." Nevertheless, the teacher can describe probable consequences of certain actions and, because basic desires and aspirations of individual human beings are quite similar, influence the judgment of his young audience.

Many of man's perplexing problems are biological ones. After all, man is a living creature and the truly serious problems involve either his continued existence, his physical comfort, or the quality of his life. Elementary biology courses, then, should serve to present biology not only as a science but also as a vehicle for discussing the wider consequences of man's various activities.

My presentation is concerned specifically with the social problems that arise from genetics or have their roots in genetic phenomena. In large part the discussion will center on races and racial differences, eugenics (including the willful tampering with genes carried by both body and germ cells), and exotic means of reproduction. Unlike my colleagues in molecular biology who have discussed the joys and sorrows of "genetic engineering" at recent meetings, I will first take up the matter of race for, unless we in the United States obtain proper perspective on racial matters quickly, the more exotic problems of genetics may be postponed indefinitely.

Biological Races

The race problem is undoubtedly The Biological Problem confronting us in America today. I say "biological" problem because the physical characteristics that serve for many persons as "racial" labels have a genetic basis. Man is a living creature; distance and the limited dispersal of individual human beings subdivide mankind into isolated populations and communities; natural selection (and chance) leads to dissimilar arrays of genetically different individuals within the various breeding groups. Therefore, the isolated and partially isolated populations come to differ one from the other. These are the evolutionary, the biological, events that underlie race formation. The anthropologist could enlarge the preceding account by discussing the development of separate cultures in isolated communities; because man has a cultural as well as a biological heredity, these differing cultural characteristics also tend to persist within populations and societies from one generation to the next.

There are certain aspects of race and race formation that should be made clear from the child's very first biology lessons. Race is a population, not an individual, concept. An individual is a member of a given race not because he has certain physical characteristics but because his parents, his grandparents, and his more remote ancestors were reproductively participating members of a population that was at least partially isolated from other populations. Black is not a race nor is white a race. Black and white are skin colors. Within any population that possesses the responsible alleles, individuals with certain genotypes will have black skins and these genotypes will appear in a certain proportion of persons in the population.

Now, gene frequencies differ considerably from population to population. Africans have darker skins than do Europeans; in respect to skin color, there is little or no overlap in the phenotypes of the two groups of persons. The frequencies of alleles responsible for dark skin color have frequencies approaching 100 per cent in one group of populations and 0 per cent in the other.

In communities where blacks and whites live side by side, intermarriage occurs, although marriages are not contracted at random with respect to skin color. Nevertheless, the composite population possesses intermediate gene frequencies in respect to skin color alleles. In such a composite population, even if marriages between light and dark skinned persons were to occur in a random fashion, a constant proportion of black skinned persons would arise generation after generation. To regard this fraction of the population as a black "race" is meaningless; those persons who by chance receive the alleles for dark skin color need not be those that for any other trait have received alleles recognized as being of African origin. The ultimate in racial foolishness was reported recently from South Africa where a dark skinned baby was removed from his parents because the authorities classified him as "coloured" while others in his family, including his parents, were classified as "white." This nonsense will be committed by any group that attempts to identify race with *individuals* rather than with *interbreeding populations of individuals*.

The main points of this section are summarized in the following four diagrams. (See pp. 76-77.) Genetic details have been sacrificed in these figures for the sake of clarity; the illustrations could, however, be easily modified to conform to Mendelian inheritance.

Figure 1 represents a community of one hundred persons, thirty of whom differ from the remaining seventy in possessing five physical characteristics each; these characteristics may be thought of as skin color, hair texture, or any of the other possible differences between men. These two groups of individuals have been drawn from non-interbreeding populations and so they fit the definition of race given earlier in this section.

Figure 2 illustrates the distribution of physical characteristics in the same community after many generations during which no barriers to the random selection of marriage partners existed. The distribution of characteristics has been obtained in the following way: The thirty members of one race differed from the remaining persons by five characteristics each; this gives a total of 150 unit characters. In distributing 150 characteristics randomly among one hundred individuals, the average number per person is 1.5. If the characteristics are distributed independently of one another and in a random fashion, a predictable number of persons will have 0, 1, 2, 3, or more characteristics each (this distribution is known as the Poisson distribution).

Figures 3 and 4 illustrate the contradictions that arise from definitions of race that are based on individual characteristics rather than on characteristics of populations. In the United States, for example, a Negro is defined as a person who has had an African among his ancestors. The corresponding definition in the hypothetical community shown in Figure 2 is that anyone showing any of the five characteristics belongs to the "black" race in Figure 1. According to this definition, eighty of the one hundred persons of the randomly mating community are "black." A spy working for the "white" race and operating with this definition of race would report to his employers that the blacks had increased in number from thirty to eighty and that the whites are now outnumbered.

Proceeding on precisely the same logic, a spy working for the original "black" race would define as "white" any person who possessed fewer than the five distinguishing characteristics. This is, in fact, the definition of Caucasian in Brazil. Of the one hundred individuals represented in Figure 2, only one possesses all five rings; in Figure 4 that individual and five others possessing four rings each have been shown as a group of six. The spy would report to his "black" employers that in this community the blacks had been virtually assimilated by the whites.

The explanations for the events that are represented in these four diagrams were discovered many years ago by Gregor Mendel; the implications of Mendelian inheritance have not yet become clear to the general population. The two groups of persons represented in Figure 1 could be referred to as belonging to different races because the ancestry of each person traced back to one or the other of two isolated populations. Once intermarriage within the community took place, the notation of two races *within* the community should be abandoned because the ancestries of different persons become intertwined. The "races" depicted in Figures 3 and 4 are not races in any meaningful sense; they are races of discrimination only. The absurdity of the definitions of race used in preparing Figures 3 and 4 becomes clear when you recall that, although the definitions are logically identical, they lead to widely different, even contradictory, conclusions.

Racial Intelligence

One of the more insidious notions accompanying the view that "black" and "white" constitute races is that of systematic genetic differences in racial intelligence; white propagandists claim, of course, that the intelligence of Negroes is inferior. This allegation should be vigorously attacked as one unsupported by sound experimental data. Past attacks have been based largely on the arguments (1) that we do not know how to measure intelligence and (2) that races do not exist. In my estimation, these arguments do not deal properly with the matter. Whether we know what they mean or not, we do collect test scores from students and we do give preferential treatment to those students whose scores are highest. Furthermore, while it is true that "black" and "white" are not biological races, persons *are* classified according to these characteristics. In claiming that the intelligence of Negroes is lower than that of whites, those making the claim are really saying that blacks obtain lower IQ scores on a variety of tests than do whites. These persons have only a peripheral interest in the true nature of biological races.

Experimental design and the interpretation of data are legitimate topics for science classes in elementary and secondary schools. By "experimental design" I mean the development of experimental procedures that yield observations bearing on the question at hand rather than on some other, unasked question.

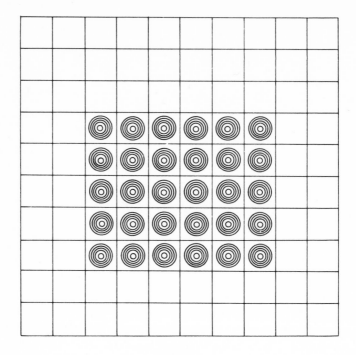

Figure 1.

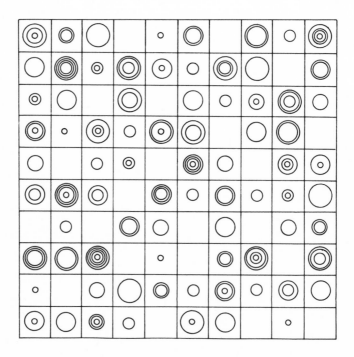

Figure 2.

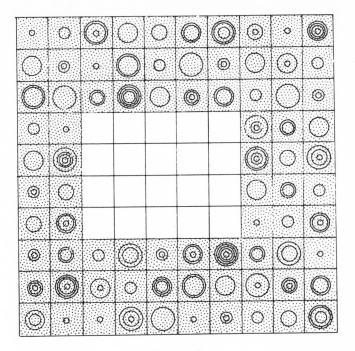

Figure 3.

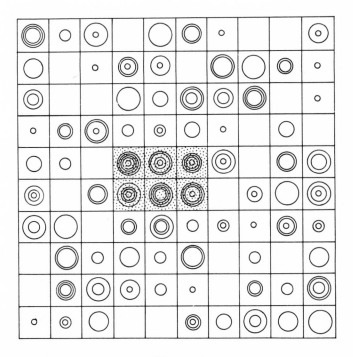

Figure 4.

Good design requires thoughtful procedures. The interpretation of observations collected by a well-designed experiment follows a predesigned scheme rather than *ad hoc* arguments or belated rationalizations. Furthermore, the abandonment of an interpretation arrived at by a thoughtfully designed experimental scheme should be regarded as a serious act calling for a complete review and the possible discarding of the questioned procedures.

The routine and repetitive use of a convenient analytical procedure is not the way to conduct an experiment. Accumulating masses of observational data by experimental procedures that are subject to criticism in no way lessens the weight of that criticism. Consequently, the compilation of results obtained from hundreds of studies on "racial intelligence" is useless if their common investigational procedure is questionable. Accumulated masses of biased data do not remove the bias; they tend, simply, to make the bias statistically significant.

In the study of racial intelligence, the classification of subjects into white, Negro, and intermediate types by the investigator repeats a classification that is made by parents, friends, neighbors, landlords, police, and many, many others in every person-to-person encounter of the subject's life. To claim that "racial intelligence" is correlated with physical characteristics such as skin color, hair texture, or facial features rather than with the accumulated personal experiences of the subject that are in turn correlated with the same characteristics is preposterous.

During the preparation of this manuscript, a lengthy article on racial intelligence appeared in the *Harvard Educational Review,* Once more, the basic error has been committed: the investigator both tests and classifies his subjects. If he can classify them, so can others. To illustrate systematic genetic differences in respect to the intelligence of races, it is necessary to test genetically segregating material. The Nobel Prize was awarded Professor Joshua Lederberg for his work on the genetics of bacteria, a science that awaited his discovery of sex in bacteria and the specially designed experiments that made this discovery possible. Professor Lederberg's numerous predecessors who studied differences between bacterial strains, differences that appeared to be characteristic of isolated strains, failed to establish that these differences were indeed genetic. They failed despite strenuous efforts because sexual crosses between differing strains and the analysis of segregating offspring are logically essential for eliminating environmental factors as an explanation for differences between isolated strains. The compelling need for a properly designed experiment has not yet struck those psychologists who insist that there are systematic "genetic" differences in respect to the intelligence of racial groups.

Eugenics

Racial differences exist because of man's evolutionary response to local selective forces and not because of any particular act of man, himself. Eugenics, the science whose aim is the betterment of man's hereditary endowment, is man-made but manages, nevertheless, to excite persons nearly as much as matters of race. Disdain is part of the reaction to eugenics; disdain because of the quaint ideals the early eugenists pushed as "desirable" and because of the trivial acts they, in their zeal, classified as "degenerate." It is no exaggeration to say that Jiggs of the comic strip "Bringing Up Father" was an undesirable degenerate by those early Victorian standards. In a more horrendous way, the reaction against eugenics stems from the demonstration by Hitler's Germany of the extremes to which a society will go in trying to achieve a desired end.

Despite the ridiculous nature of some early claims and the terrifying procedures adopted by the Nazis, selection against certain mutant genes does go on in human populations and, by the use of his genetic knowledge, man can make this selection more efficient (and less painful) than it is at the moment. Hemophilia, for example, is a rare human trait caused by a defective sex-linked recessive allele. The blood of hemophiliacs is unable to clot; severe bleeding accompanies the slightest cut or scratch. Hemophiliacs, almost always male, lead a miserable existence; if they do not die at an early age, they are nevertheless crippled by painful hemorrhages and swellings in their joints. Female carriers are unaffected by the disease but one-half of their sons will be hemophiliacs and one-half of their daughters will, like their mothers, be carriers of the mutant gene.

Knowing from past experience the inheritance of hemophilia in human pedigrees, what options are available in today's society? First, we might remain aloof while allowing events to take their "natural" course. We can, for example, allow a woman who has borne one hemophiliac son to bear additional children without sharing our knowledge with her. That, I believe we would all agree, would be an utterly reprehensible attitude. Man increases his store of knowledge not to create a clique of eye-winking, knowledgeable, but silent spectators; on the contrary, he strives to increase the level of knowledge for all persons.

A second possibility might be to warn the carrier of the hemophilia gene that one-half of her sons will be hemophiliacs like the one already born. That warning, however, is itself a eugenic measure! The woman may decide to forego any additional children. On the other hand, either by accident or design, she may become pregnant once more. Again, we might adopt the attitude, "Let's wait and see what the child is like." But, it is not necessary to wait nor do I think that it is *right* to wait.

The sex of very young unborn embryos can be determined by obtaining a sample of cells from the fluid surrounding them in the womb. If an examination of these cells shows, in the example we have been citing, that the unborn child is female, she will not suffer from hemophilia (although she may carry the defective gene); if the test reveals that the baby will be male, then it has a 50:50 chance of being hemophiliac. This would be sufficient cause in my estimation to warrant a therapeutic abortion. On the other hand, subtle tests may eventually be developed that will reveal whether the unborn male fetus is in fact afflicted with hemophilia. I can see no reasonable alternative to the early abortion of a male hemophiliac embryo.

Genetic Engineering

In years to come, eugenics promises to take forms quite unlike those of the past. The bulk of present day eugenic measures consists of encouraging childbirth in one set of parents and discouraging it in another. The nature of the means for "discouraging" determines the emotional reaction of the population to eugenic measures; so far we have witnessed deterrents to reproduction in the form of oral persuasion, mandatory sterilization, and, in Hitler's Germany, legalized murder.

A series of procedures has been suggested by molecular biologists for the relief or cure of genetic disorders. One of these procedures (euphenics) calls for the supply of missing substances to genetically handicapped persons. The injection of insulin into diabetics is a euphenic measure; it is a procedure that corrects the phenotype of the genetically handicapped without correcting the underlying genetic defect. Similarly, omitting phenylalanine from the diet of babies suffering from phenylketonuria and of milk from those without lactase are euphenic measures.

More sophisticated euphenic procedures might involve not only the addition or deletion of relatively simple metabolic products from the diet but also the use of specific gene repressors or activators for those abnormalities where the control of gene action (rather than the nature of the gene product) is at fault. Nevertheless, the aim of such therapy would still be the alleviation of symptoms; the underlying defect would still be present to be passed on to succeeding generations.

Genetic engineering is a second procedure that has been suggested for the cure of genetic ills; it consists of the incorporation of normal genes within the protein coats of human viruses and the introduction of those genes into somatic cells by infection. Analogous procedures are used routinely in bacterial genetic studies. The procedure is not an impossible one for human beings as one group of laboratory workers has already demonstrated quite by accident. This chance demonstration involved doctors and technicians working with the Shope papilloma virus, a virus that induces tumors in rabbits but which is non-pathogenic in man. Although the virus is non-pathogenic to humans, it appears to infect many persons who handle it and at least one of the viral genes, the one responsible for the synthesis of an easily recognized arginase, is active within human cells. Infected laboratory workers, even years after their last contact with the virus, have especially low levels of arginine in their blood as a result of the viral enzyme. As Professor Lederberg has commented, it is unfortunate that a disease characterized by an excess of arginine does not exist in man for, if it did, we would now have discovered a specific remedy, the first medical triumph of genetic engineering.

A third procedure calls for the replacement of mutant genes by normal ones through the treatment of germ cells prior to fertilization; this remedial action would persist within the population because children and later descendants of the originally defective person would be normal in respect to this particular defect.

The possibility for carrying out these, and still other, genetic tricks certainly exists. Whether they will in fact be implemented in the near future is problematical. Mammals are much more complicated than bacteria. The regulation of gene action is a complex problem in a higher organism and it is one that in the past biochemists have consistently underestimated. Not long ago we were told that the active site of an enzyme was the only essential part of the enzyme molecule; the rest of the protein molecule was said to be made by "fossil DNA." More recently, allosteric properties of enzymes have been discovered; these refer to the change in the configuration of enzyme molecules when they combine with certain small molecules that act as enzyme activators or inactivators. Allosteric changes represent a form of enzyme regulation. In my opinion, problems of communication and regulation within developing higher animals multiply much more rapidly than the obvious physical complexity of the animal itself. Thus, I wonder about the success of some of the postulated euphenic measures.

Quite apart from the matter of success, there are divergent views regarding the desirability of these engineered changes. Doctor Nirenberg has urged that we refrain from such tamperings until we can see precisely what we are doing; Lederberg has spoken out for experimentation whenever experimentation appears to be called for. Indeed, as Lederberg has urged, a physician might be reprimanded for *not* attempting a promising new treatment if other remedial procedures have failed in a given instance.

If I force myself to think small, I find that I favor Lederberg's point of view. I am particularly impressed by the individualistic nature of the experimental state—one patient, one doctor—that Lederberg emphasizes. To delay the early experimental trials while investing tremendous sums of money in a given remedial procedure causes pressures (face-saving, among other kinds) for its large-scale use. And so, if an error in judgment has been made, it will have an enormous impact on the population. This is the outcome I fear when I find myself thinking big. As I mentioned earlier, we must now wait for the final effect (if any) of the simian virus widely dispersed in man by way of polio vaccinations. I would rather risk small scale successes and small-scale errors than large scale successes and large-scale errors. Success can be amplified; a blunder, once committed, cannot be retracted.

The specter that haunts many minds is one in which genetic control of citizens is taken over by the government. Anything is possible, of course, but I prefer to think that citizens will not take leave of their senses en masse. If I did not believe this, there would be remarkably little in life that would cheer me. If I did not believe that the government represents people and that the majority of persons are sane and reasonable, I could become terrified at the thought of artificial insemination, lack of gun controls, compulsory military training, or income taxes. Several of these items do depress me, to be sure, but I have never really expected the government to take *all* of my income, or to make me serve my *entire* life in the army. By the same token, I do not expect the government to remake its citizens genetically according to some mysterious, diabolic plan.

There remains, however, a problem that is raised by euphenic measures. Individuals carrying defective genes are placed in a position—simply by surviving where in earlier times they would have died—from which defective genes can be passed on from generation to generation within the population. Should we worry about this? Should we permit it? The obvious answer is the algebraic one provided by population genetics: the frequency of a mutant gene increases to a level commensurate with its deleterious effect on its carriers. Hence, if euphenic measures mask all deleterious effects of a gene, there is no harm should its frequency reach 100 per cent. This line of reasoning assumes, however, that populations approach new equilibrium frequencies smoothly. I have been impressed recently with the irregularity of population changes. The Potato Famine in Ireland during the 1840's represents either a zig or a zag on the curve representing Ireland's approach to an equilibrium population; 2.5 million of 9 million persons either starved or emigrated during this "irregularity." Catastrophes of this sort should be avoided. Unforeseen events might interrupt an otherwise standard euphenic treatment to a large proportion of a population and thus lead to a large-scale catastrophe. To avoid this possibility, I would prefer that persons saved by euphenic measures be discouraged from having children. This potentiality need not be traumatic because in the very near future some sort of population control will be required and many persons on that account will be childless.

The Clonal Growth of Human Beings

The asexual reproduction of man is still another possibility that exists, thanks to modern biology. Asexual reproduction in man is not really new; each pair of identical twins is produced from a fertilized egg that starts life as one individual. At some point in the early development of an occasional zygote, it divides amoeba-style so that two separate cells, or small clusters of cells, eventually produce two separate individuals. The man-made possiblity, however, involves many, many eggs from which nuclei are removed and into each of which one of a series of identical nuclei is introduced. The resultant array of persons would be as genetically identical as are identical twins.

The social implications of clonal reproduction revolve largely around the mechanics of child production, not around genetic matters. What is the market for children produced artificially? Who determines the number that will be produced in a world that is already over-populated? Will governments attempt to swell the number of engineers, or professors, or soldiers in this manner? To be frank, I am against it. The truly large problem of the world has two components: the first is to lower the numbers of persons. The second is to raise our technological machinery at the service of each to levels compatible with man's continued existence on earth in physical comfort and in harmony with himself and his natural surroundings. The attainment of these goals will be difficult enough with old-fashioned reproductive techniques; I would certainly oppose the licensing of a "General Mills" for clonal persons.

The possibility of using nuclear transplants for the production of genetically identical persons raises the question, "Who am I?" and "Who are you?" One father whose son was entering military service spoke recently of maintaining a tissue culture of his son's cells so that, in the case of the boy's death, he could be reconstructed once more from the tissue culture. This attitude reveals a profound ignorance of what makes a person a person. The all-important feature of any person is his mind. It is in the mind that the experiences of a lifetime are recorded. To assume that parents can duplicate these experiences identically after a lapse of twenty years in order to reconstruct the same son once more is pathetic beyond belief.

(Postscript added later: Were I to rewrite the section on clonal reproduction today, following many recent discussions with students on contraception, abortion, and population control, I would be inclined to emphasize the following: The possibility of obtaining hundreds or even thousands of identical embryos from cells scraped, for example, from the ephithelial lining of one's mouth—cells that otherwise would slough off and die—should call for a re-evaluation of some of the taboos against abortion. The familiar landmarks (fertilization, or union of sperm and egg pronuclei) normally associated with reproduction and the creation of a new life are now missing. The impact of clonal reproduction may be more important in respect to our definition of what is an individual and, hence, to our attitude toward abortion than in respect to the possible wholesale production of individuals by a new reproductive technique.)

CHAPTER FOUR ✳ GENETICS ✳ *The Genetics of Racial Differences in Intelligence* by JAMES C. KING

I would like to devote the space available to me to a bit more detail on a subject mentioned by Dr. Wallace: differences between races in intelligence. Professor A. R. Jensen's recent paper, which has had very wide publicity, has, I am afraid, raised doubts in many minds that any substantial increase in IQ can be brought about by environmental changes. Let us see just what Jensen's paper has to say on this point.

Jensen cites statistical studies showing correlations between the IQ's of persons of known degrees of relationship—the most straightforward cases being based on pairs of monozygotic twins raised apart. These studies give an estimate of the heritability of the IQ of .8. Heritability is defined as the proportion of total phenotypic variance in a population attributable to genetic differences. One minus the heritability, then, represents the proportion of the total variance attributable to nongenetic factors. Hence environmental differences cannot possibly have contributed more than .2 of the total variance. Then Jensen says, ". . . on the average, Negroes test about one standard deviation (15 IQ points) below the average of the white population." He continues, ". . . various lines of evidence viewed altogether, make it a not unreasonable hypothesis that genetic factors are strongly implicated in the average Negro-white intelligence difference." He infers that since nongenetic factors contribute so small a portion of the variance of the IQ and since there is so large a difference between the mean IQ's of white and Negroes, changes in the environment are not likely to bring the two means much closer together.

Heritability is a concept which has been worked out carefully by mathematical geneticists and it has been found very useful by plant and animal breeders for predicting the results of breeding programs. It is a characteristic not of an individual nor of a trait but of a population. Jensen himself recognizes this. "Estimates of H," he writes, "are specific to the population sampled, the point in time, how the measurements were made, and the particular test used to obtain the measurements." One minus H for a given investigation does not at all represent the possible latitude of environmental influence on intelligence; it merely tells us how much, in that experimental situation, the total variance would have been reduced if all environmental differences had been eliminated. Under other environments, entirely different values for total variance, for H and for one minus H may occur.

Again quoting Jensen: "All the major heritability studies reported in the literature are based on samples of white European and North American populations, and our knowledge of the heritability of intelligence in different racial and cultural groups within these populations is nil. For example, no adequate heritability studies have been based on samples of the Negro population of the United States." In other words, we have some evidence that H is .8 for the white populations of Britain and the United States. We know nothing about the value of H for Negroes. Yet the figure .8 was used in deducing that environmental changes probably could do little to raise the mean IQ of the American Negro population.

In discussing the effects of environment on intelligence, Jensen says: "Below a certain threshold of environmental adequacy, deprivation can have a markedly depressing effect on intelligence.

But above this threshold, environmental variations cause relatively small differences in intelligence. The fact that the vast majority of the populations sampled in studies of the heritability of intelligence are above this threshold. . . accounts for the high values of the heritability estimates and the relatively small proportion of the IQ variance attributable to environmental influences." In other words, the figure .2 as the environmental component in the variance of the IQ is so low because the population samples on which it was based were characterized by small differences in environment.

Jensen goes on to say, "The environment with respect to intelligence is thus analogous to nutrition with respect to stature. If there are great nutritional lacks, growth is stunted, but above a certain level of nutritional adequacy . . . even great variation in eating habits will have negligible effects on persons' stature, and under such conditions most of the differences in stature among individuals will be due to heredity."

It is interesting to note that in the group of monozygotic twins, raised apart, which furnished the estimate of .8 for the heritability of the IQ, the heritability for stature was .86. This would give an estimate of .14 for one minus H for stature and make an increase in stature as a result of environmental change even less likely than a rise in average IQ. Yet, for the past several decades the mean stature in Western Europe, the United States, and Japan has been increasing by startling amounts. In the United States from 1918 to 1958 the mean stature of adult males rose by very nearly one-half of a standard deviation. There is little if any evidence to indicate that this increase is the result of genetic change. It must be the result of as yet unanalyzed changes in the environment. There is no evidence in Jensen's data, or in any other body of data, to suggest that a similar increase in the IQ might not occur.

CHAPTER FOUR ✳ GENETICS ✳ *Genetics, Behavior, and Social Responsibility* by RALPH HILLMAN

Recently Warren Weaver wrote that the morality problems which arise in the search for new basic knowledge are becoming more and more obvious to both scientists and laymen. Part of Weaver's thesis was that the products of basic research are powerful tools in the hands of society, but that it is foolhardy to believe that these products invariably accomplish good. The rich resource of new knowledge which comes from basic science, therefore, creates a responsibility both for those who produce and for those who support the production of this knowledge. This responsibility is the assurance that the use of this knowledge serves the society as a whole.

I think that we can all agree that it would be very difficult, if not impossible, to stop basic research. Man is a curious animal, in both senses of the phrase, and he has both the physical and mental capacities to satisfy a great part of his curiosity. You really could not stop him. It is not, however, basic research and discovery that has begun to frighten a large segment of our population. It is rather the fact that man has a horrible propensity to misuse that which he has discovered. This misuse, moreover, is not limited to natural resources but also includes mental resources and ideas. What in fact terrifies some people is that in our modern society, where ideas and technology are caught up in a "space-race" philosophy, the limiting margin for error in the use of knowledge is rapidly shrinking.

We can look, for example, at the problems which might arise when uncontrolled population growth takes place on a planet with finite resources. How does a society deal with the overcrowding which comes as a result of an expanding population? Upon what basis will the rules of population control be formulated? What will be the model for population control, and who will determine this model? What will be the decision-making process, and how will genetics influence these decisions?

Our present knowledge of two genetic diseases, hemophilia and diabetes, one presently incurable, the second able to be kept under control, may be used to illustrate the problems which mankind may soon have to face. Should we treat the hemophiliac by abortion and the diabetic by insulin? Should we attempt to cure hemophilia or should we prevent the individual who has this disease from passing his genes on to future generations? If this question isn't difficult enough to answer, let me ask still another. Huntington's Chorea is a genetic disease caused by a dominant gene, which strikes fatally after the reproductive years are over. In a time of limited research resources, in terms of both men and money, coupled with the pressures of overpopulation, should the effort be placed on finding a cure for this disease or should it be placed on finding definite methods of heterozygote detection so that reproduction by those heterozygotes will be controlled? Should we in fact spend our limited resources to cure a disease or to remove a "genetic load" from the population? Where does man's altruism stop and the rationalization of his actions begin?

These questions do not involve knowledge, and the use of knowledge, as much as they involve man's understanding of ethics and ethical behavior. Robert Sinsheimer recently wrote that the knowledge for the designed improvement of man is present in our understanding of transformation and transduction and that all we need today is the technology. The question is really what path will this technology take. Will it take the short-range path of self-esteem, or will it look toward the search for humanism and humanistic behavior?

The future of man on this planet may well lie in a knowledge of the controlling elements in his behavior. The role of genetics is obvious. The imperative is to determine the inheritance of factors which control behavior in man and to explore the mechanisms underlying these factors. Roger Williams recently stresssed the point of view that heredity must be used to help us to understand and solve the human problems which have been the province of the social sciences. The social significance of genetics is certainly related to investigations of the hereditary control of human behavior on an individual and group basis.

CHAPTER FOUR ✳ GENETICS ✳ *Can Research Help us Read Hereditary Differences in Nervous Systems?* by E. MARIE BOYLE

Dr. Wallace's presentation leads me to speculate about a number of biological problems that are related to the field of genetics. Certainly in today's modern high school classroom there should be serious consideration of the implications of current research.

This need not become sheer aimless wandering in the distant areas of science fiction. Concerted group effort, as a class tries to use its own newly acquired understanding of biological principles to see how relevant problems are being investigated today, can give the teenage class member the kind of practice required by the voting citizen as he evaluates programs proposed for government action.

The designing of experiments and the interpretation of resulting data demand clearer thought in the whole area of race and of culture, as Dr. Wallace has indicated. Additional examples of valid investigations would help many teachers as they confront classroom misunderstandings.

Eugenics, euphenics and genetic engineering hold great potential for exploration and development in the future, but the general question that applies here is: *who* shall decide on the use of the knowledge we have and that we will discover? Dr. Wallace says, "I prefer to think that citizens will not take leave of their senses en masse." I believe in education and therefore I also believe with him that this can be prevented, but it *has* occurred —as, for instance, in Hitler's Germany. It is up to us, in the teaching field, to insist on independence of thought and its importance by our *everyday support* of those who *do* have the initiative to ask the question that is different or to pose the possibility that may seem "far-fetched."

The last few closing statements of Dr. Wallace's paper raise many questions in my mind. In mentioning tissue culture as a means of producing identical persons, he introduces the problem of what makes a person a person and refers to the mind as the "all-important feature of any person" wherein "the experiences of a lifetime are recorded." This seems to me to ignore the whole significance of values and the processes by which mind, emotion, and experience interact. Note, for instance, the way many excellent minds are differing so widely as to the current anti-ballistic missile issue.

The study of values and of the bases for behavior is, of course, the field of behavioral psychology, and I am well aware that this field of psychology is a complex one. But complexity is not an acceptable reason for failure to carry out scientific investigations. I feel we need to know more about our *inherited* neural characteristics so that we can be more effective in assisting each individual to develop his highest potential. Every parent knows intuitively that any two of his children came into this world with differing nervous systems. The kind of handling that "works" for one new-born baby does not do nearly so well for the next. And this is not *just* a result of the experience or the relaxation of the tension on the part of the parent. It is also an inbuilt (which is to say, hereditary) characteristic of the child. What actual physical bases go to make up this difference in response? How can we learn to differentiate among types of inherited nervous systems?

What measures can be developed to demonstrate inherited neural differences? We know a little, I gather, about variations in speed of reactivity, but what lies behind these? Is it a chemical variation that allows for speedier synapse action or is it a function of the central brain "switching board" reaction? What degree of reliability can there be in new types of measurement in this field? Can some connections be made between these physical measurements and basic behavior patterns?

We are gaining some information about basic patterns of sensitivity correlated with circadian rhythms. Do individual differences in reaction to anaesthesia, pain-killing drugs or poisons, which may be related to these biological clocks, give us any clues that might lead to the establishment of basic patterns of sensitivity in general? The amount of information delivered to us by our senses can be increased by training which we could call: increasing our sensitivity. But I am referring to a much more deeply-seated, body-wide, inherited kind of sensitivity.

Let's not restrict our investigations in this area of the combined disciplines of genetics and behavioral psychology to rats or to primates other than ourselves. Much more than we presently know can be discovered about man by examining human beings without harming individuals. My hope for the possibilities in this field was much increased when I recently came across a review in *Bioscience* of a book edited by Jerry Hirsch. I gather from the reviewer that "further work needs to be done" because of "the distressing inadequacy of basic information." But at least the combined field of research *is* in existence. When we have learned more about the inherited neural bases for our behavior, just think of the possibilities of the use of this information about types of inherited nervous systems.

Parents would be able to "cope" much more happily and successfully with *each* of their children. Education could at last be meaningfully individualized. In other words, cultural acquisition could be less painful and not so restrictive to the individual. This could lead to mature citizens who could adventure freely in giving culture new directions of real worth. Our world *could* be, and let us all work so that it *will* be, an infinitely better one!

CHAPTER FOUR ✳ *Discussion*

QUESTION—*Please discuss the question of genetic differences between races in a little more detail.*

DR. WALLACE: Suppose one has two strains of some organism—two cultures of bacteria, for example. One does not demonstrate a genetic difference between these strains by simply noting that they differ. To demonstrate a genetic difference, one crosses individuals of the two strains and examines the progeny of the cross. Many early bacterial geneticists knew that bacterial strains differed; these persons wanted very much to say that these differences were genetic but they could not make that claim until Joshua Lederberg discovered bacteria that exchanged genetic material.

In moving on to a higher organism, I will need to explain a term: mid-parent. If the two parents in a cross of fruit flies, for example, differ in some measurable trait, the values of these two parents can be added and then divided by two. The average obtained in this way is the mid-parental value, a value that can be compared with the average value of their progeny. In a culture of flies that has been maintained in the laboratory for some time and which is not evolving, the average value of the parental generation equals the average value of the succeeding one.

Suppose now that I construct a graph on which I plot (vertical axis) the mid-parent values for individual crosses *versus* the mid-values for their progeny (horizontal axis). I will most likely get a series of points that will fall in an oval that slopes from lower left to upper right; the slope of this oval is reliable evidence (but not proof) that the trait under consideration has genetic basis.

Now, suppose I have two stocks of flies that have been maintained separately for a considerable time. Suppose, too, that upon plotting mid-parent *versus* mid-progeny values each stock gives an oval diagram that slopes from lower left to upper right. Suppose further, however, that the mid-parent-mid-progeny point of one strain is displaced downward and to the left relative to that of the other. Does this displacement mean that the two strains differ genetically in respect to the trait we are discussing? No. That the trait in question has a genetic basis within each of two stocks of flies does not mean that the difference between the stocks reflects a genetic difference between the stocks. Indeed, if environmental differences exist in the maintenance of the two stocks, the observed difference may lie in quite the opposite direction from any systematic genetic difference that may exist.

Now, there are techniques by which hidden genetic differences can be revealed free of environmental influence. These techniques work very well with lower organisms. Their use in human genetics tends to be confused by man's "cultural heredity"—that is, by the influence of parents' status in society on the fate of their children. They are techniques that would prove to be tremendously expensive if applied to man. I am not convinced that a study utilizing these techniques is worth doing. Nevertheless, the data now available that purport to show systematic genetic differences between blacks and whites in respect to IQ do nothing of the sort; the available data are not the sort that yield genetic proof.

DR. KING: The set of experiments that Dr. Wallace outlines is impeccable from the point of view of the geneticist but it might be difficult to get across to the general public. In discussing this question it should be emphasized that all the arguments which contend that a genetic difference in intelligence has been demonstrated depend on the fact that, in general, the mean IQ of Negroes tested is fifteen points lower than the mean IQ of whites tested. This is presumed to be evidence of a genetic difference. The tacit assumption is that the environmental situation is the same in both cases. This implies that in some cases we have raised whites in a Negro environment and in other cases we have raised Negroes in a white environment. As a matter of fact the "Negro environment" is based upon the classification of a person as a Negro, consequently it is impossible to raise a Negro in a white environment in this country, just as it is impossible to raise a white in a Negro environment. Therefore, the argument that we have demonstrated genetic differences by comparing the IQ's of individuals raised under the same environment simply is not valid.

DR. WALLACE: Dr. King refers to my proposed experiment as "impeccable." Actually it is not and therefore a manuscript I once wrote in which it is described has never been sent off for publication. The test I described, because it cannot eliminate the effect of "cultural inheritance" that I mentioned earlier, still penalizes blacks. If I have some notion of how a test might turn out if it were to be a perfect test, I might settle for a less-than-perfect one if it introduced unwanted biases that oppose my expectation. I would not perform the less-than-perfect test if its biases were to reinforce my expectation because I would not be able to interpret the data if they happened to support my views; there would always be the possibility that my expectations were wrong and only experimental bias has made the data appear to support them.

I want to repeat that the question of a genetic difference between races in respect to IQ should really be phrased: Does DNA of African origin lower the IQ of its carriers relative to DNA of European origin? And I submit that this question cannot be answered by administering IQ tests to members of isolated groups.

QUESTION—*I believe that Dr. Jensen's article points out that the blacks in other countries have a superior IQ with an inferior environment.*

DR. WALLACE: I've a feeling that will take a psychologist to unravel. I speak as a geneticist and my contention is that when you say that something has a genetic basis you refer back to the DNA molecule. The demonstration of a genetic basis requires a breeding experiment; it cannot be demonstrated by making a comparison of isolated groups.

QUESTION—*Please comment on Miss Boyle's statement about what makes a person a person.*

DR. WALLACE: Miss Boyle's statement causes me to emphasize the attributes which go to make up a human being as a human being. The important thing is the mind. Legs can be lost and replaced by artificial legs and hearts can be transplanted. Conceivably everything could be exchanged except that the person we know as a person depends on what is in his head. If we ever get to the stage at which we transplant heads, we'll have to have a rule which says the person goes where the head goes.

I want to reassure Miss Boyle concerning my statement that the mind is the all-important feature of a person. In using the word "mind," I mean to include attributes such as values and emotions as well as the mere catalogue of past experiences; we have no obvious quarrel on this point.

QUESTION—*Which techniques do you recommend we use in our classrooms to discuss these issues with high school students in a meaningful way?*

DR. WALLACE: I do not know how one talks to fifteen-year-old students; I talk primarily to undergraduate and graduate students. I find myself stressing to these students that, before running off to do something, they had better go first to the library and arm themselves with facts. Now, in the case of high school students I do not know what the best procedure would be. I suggest that you go over with the student the types of evidence that are required to justify certain claims. I also suggest that from the start students be taught that race is a population concept, not an individual one.

MISS BOYLE: Perhaps special student reports, student forums and debates might be useful, but I agree with Dr. Wallace: let's be sure it all has a factual basis.

QUESTION—*Can you get high school students to discuss race in an unprejudiced manner?*

MISS BOYLE: In my experience, this is definitely possible. In fact many young people today are skilled in discussing such questions without prejudice. They are practicing constantly and can participate in a rational way.

CHAPTER FIVE ✳ POPULATION

Can Teachers Tell the Truth About Population? by GARRETT HARDIN

In spite of a vast mass of fine rhetoric, the principal purpose of education has always been indoctrination, the instructing of young persons in such a way that (it is hoped) they will later maintain the social machinery that protects and supports the more influential members of society who pay the bill for education. How could it be otherwise? Is it not Utopian to expect any society, however idealistic, to finance an educational system dedicated to its own destruction?

Most of the time, in most places, most of the educational establishment has been paid to support the Establishment. But no system is perfect (least of all an educational system!) and, as we all know, from time to time educators have appeared on the scene and explicitly have told their charges that, contrary to what they might have heard before, they do not live in the best of all possible worlds. More significantly, some educators have even indicated that it is possible to make improvements. Improvements will inevitably be viewed as subversive of the existing order by those who prosper from the present social system. If what he says receives sufficient public notice, the erring teacher is put under pressure to censor his statements; in extreme cases he is liquidated. The classical instance is Socrates; we must not forget the hemlock.

But there have not been many like Socrates. Furthermore, society has made a sort of adjustment to the destructive observations of thinkers and teachers. A distinction has been made between ethical and practical matters. Criticisms of society can by quite effectively "denatured" by assigning them to the ethical realm, thus implying that they are Utopian in character. Practical

97

men, after making a slight note of the ethical statements, go on to make "practical" decisions in a "real world" almost as if they had not heard the ethical discussions. Many of the most profound criticisms of society have been tolerated because ethics was thus safely isolated from practice. Schooling, particularly at the college level, has been regarded as a period during which the student is allowed to hear subversive ethical discussions in the calm assurance that his passionate ethical concern would constitute only a passing phase in his development. Once out of college (his parents assured themselves) he would soon get over the wild ideas he had picked up from his professors and settle down to being "sensible."

Such has been the basis of the truce between society and those who pursue truth as a vocation. It begins to look as though the truce is at an end. For this change many reasons can be adduced. Important among these is the oft-noted fact that technological progress has outstripped social progress. Scientific and technological progress is clearly a phenomenon that enjoys "positive feedback" resulting in exponential growth at such a rate that the doubling time is in the neighborhood of ten years. It is not obvious how one would measure the doubling time for social progress, but it is indubitable that such progress occurs at a far slower rate. The discrepancy between the two kinds of progress has existed for a long time, creating great stresses that can perhaps be relieved only by something resembling earthquakes in the body politic. Let's hope not; but we do have good cause for fear.

The stresses have reached such a level that even conservative educators who accept, as an important function of the process of education, the maintenance of the stability of society are now beginning to wonder if they must not adopt some of the measures espoused by the radicals. When stability is thus endangered, it may be the better part of wisdom to deliberately relieve some stresses before they become extreme. It is such a consideration that leads even a conservative educator to become a limited revolutionary.

The entire problem of revolution by education is so large that I cannot hope to illuminate it in its entirety. I will take up here only one phase of this problem, the part that is concerned with population control. After about twenty years of "viewing with alarm," population professionals have now moved toward considering what we can do about it. As to wise actions, opinions differ; but it is quite clear that no single remedial action can be taken without undermining the existing arrangements of society. Every population professional who contemplates action is necessarily a revolutionary.

Being a revolutionary is hazardous. If you tell all the truth that you know in one great blast you will either be liquidated (if you have some power) or (if you do not) "sent to Coventry," i.e., be permanently labeled as a "kook." In either event, you have lost. On the other hand, if you espouse a program of "gradualism," the rate of progress you encourage may lag behind the rate at which the problems become worse. There is the dilemma.

There is no overall solution to this dilemma. We simply must, somehow or other, find an acceptable middling course. I doubt if there is a single best course. Solving the action equation for its optimum we must plug into it parameters derived from the local scene in which each of us works. Solutions are local, both in time and place. The best *general* thing to do, I think, is to publish a *Graduated Checklist of Heresies* which each local man can use in evaluating his own efforts. Such a list I present here.

The first item on the list is so widely accepted now (though it was not even as late as ten years ago) that I think any teacher can safely tell it to his students. It is included here as a heresy only for the sake of completeness. The heretical character of the items on the list increases step by step until, with the last member, we reach an item that is so heretical that I cannot even tell of it, even to this audience—I must only hint at it. Each teacher viewing his own situation should work out for himself how far down on the list he can safely go this year. Next year, hopefully, he can proceed a few stages further. But at no time can he go all the way, because this list is open-ended at the bottom and should be explicitly extended when the present roster of recommendations has nearly all been converted to practice. For convenience I divide the heresies into four groups.

First Group

1. There *is* a population problem. No one (except for a few doctrinaire Marxists) doubts this now. Yet only five years ago a prestigious committee of the National Academy of Sciences felt obliged to labor hard and long in producing a wordy report that boiled down to just this one conclusion. The mountain labored and brought forth a mouse. But perhaps it was necessary.

2. Our population problem is an evil byproduct of the conquest of disease. If Louis Pasteur (and all he symbolizes) had not existed, we would not be bothered with overpopulation. Crowd diseases would periodically solve our problem.

3. War is no solution. The population problem is essentially an imbalance between resources and numbers of people. With modern weapons we can kill people by the hundreds of millions—but modern weapons are more destructive of resources and organization than they are of human beings. Each time they are used the population problem becomes worse, not better. (Biological warfare possibly offers us a technology that would improve the population situation. The "improvement" is one that few reformers would recommend.)

4. As regards population growth we are living in very exceptional—perhaps we should even call them abnormal—times. At the present time the world is increasing at 2 per cent per annum. If this had always been the rate of increase, Adam and Eve would have been born only ten centuries ago. If population continues increasing at this rate, there will be "standing room only" in a little more than six centuries from now.

5. The world is limited. Its mass is only 5.983 times 10^{27} grams. There is no reason to think that all of this mass can be converted into human flesh, even if this were desirable.

6. "Space" is no escape from our earthly population problem. The reasons for this statement are many, complex, and completely convincing.

7. The limit of human population depends on what you assume (or want to be) the limiting factor. If nothing else, heat dissipation will set a limit to terrestrial population. This limit—which we may call Fremlin's limit—is 10^{18} or a billion billion people. Those who have read J. H. Fremlin's essay understand that life would be all but unbearable in the near neighborhood of this limit.

8. Zero population growth must soon be accepted as the normal state of affairs. It has been the normal state for most of human history; it will soon again become normal.

9. The maximum population is not the optimum population. The proof is quite simple. Energy, on which life depends can be measured in calories. Our daily caloric needs can be divided into existence calories and enjoyment calories. Existence calories are about 1,600 calories per day—the minimum on which we can just

exist, unclothed, unhoused, unentertained: just lying down and existing. Enjoyment calories are used for all other purposes: housing, clothing, furniture, athletics, travel, dancing and operating all our machinery. In the United States, the enjoyment calorie budget is fifty times the existence calorie budget. To maximize population we would have to reduce the enjoyment calorie budget to nearly zero. But who would regard such a style of life as ideal, or such a population as optimum?

Second Group

10. "Progress" is the opiate of the people. The "Idea of Progress" is essentially a religious idea. What it implies is that technology can do no wrong. But if we are to survive we must establish the idea that mankind is under no compulsion to use every invention that we make. We have no religious obligation to build every building that can be built, to pave over the entire country with highways, or to destroy all natural beauty in the name of efficiency. One can argue, of course, that we are not really against *progress* but only against unintelligent "Progress." This is true, but this may not be the best way to put the matter. It may be best to be blunt and irritating so as to call attention to the revolutionary character of our rejection of traditional progress. Before people can achieve insight they must be psychologically disturbed. Tact is the enemy of insight.

11. Birth control does not equal population control. Without the guidance of standards, measurement, and conscious control, people will naturally breed until the proportion of enjoyment calories per person is very low indeed. There is no automatic adjustment that endures a high level of enjoyment calories.

12. If each herdsman sharing a common pasture is free to increase his herd as he sees fit, the commons is inevitably ruined by overgrazing. Freedom in a commons brings ruin to all. If human beings live in something like a "welfare state," in which everyone has a guaranteed annual wage, and if freedom to breed is not restricted, breeding will not come to an end until sheer misery acts as the controlling factor.

13. Population control cannot be achieved by an appeal to conscience. People vary in the extent and power of their consciences. If we leave breeding up to the individual conscience, we insure that those people whose consciences tell them to have

many children will have more children than the others, and will leave more descendants in each succeeding generation. The ultimate result will be a complete elimination of those consciences that speak for a small family.

14. Population control requires coercion by some means or other. Coercion need not be mechanical and rigid; it may be indirect and even subtle, e.g., by way of punitive taxes, or rewards given for desired behavior. Mutual coercion mutually agreed upon will be required before population growth can be controlled. Whether one wishes to call the control method "coercion" or not is a matter of taste. Again, I would recommend bluntness.

15. In population matters as in all others, it is true (as Hegel said) that "Freedom is the recognition of necessity." After the first shock, we accepted laws that prevented us from robbing banks, that forbade us to empty chamber pots out into the street, and that prevented us from opening up a tannery in a residential district. After the first shock, we found that such restrictive laws did not significantly diminish our freedom. In a populous world, in fact, wisely restrictive laws increase our freedom. The social control of individual breeding will ultimately (and probably very soon) be accepted as a necessity and hence compatible with freedom.

16. At the population level, food is not the remedy for famine. At the individual level it is, of course. But with respect to entire populations, when we save the people from starvation by sending them food, we increase the number of breeders and consequently increase the magnitude and tragedy of the famine that will afflict the same population a few years or a generation later. Out of kindness, we should never send food to a starving country unless we also incorporate with this aid an effective program of birth control.

17. Including food in foreign aid should always be viewed as an evil act, unless it can be shown that it will diminish suffering. The burden of proof should be on those who recommend this course of action.

18. Nonselective immigration is indefensible. The problem of population created by a country that breeds too much cannot be solved by allowing some of their people to migrate to another and less crowded country. Such migration results not in a sharing of the wealth, but a sharing of poverty. It transplants and magnifies

the problems of overbreeding. *Selective* migration is another problem. From the point of view of the national interest of the receiving country, something can be said in favor of accepting highly intelligent, highly trained, or otherwise highly desirable immigrants. Of course such a "brain drain" or "quality drain" weakens the donor country.

Third Group

19. Even over the short term, conscience is a dangerous control to invoke. Asking a person to refrain from taking goods out of a common store for the sake of society as a whole, when such a request is not accompanied by coercive laws, entraps the individual psyche in a pathogenic "double bind." He senses that if he restrains himself he will be thought a fool by those who do not; on the other hand, if he "gets his before the pigs come" he will be thought a scoundrel. In such a double bind he loses no matter what he does. Resenting this, he becomes a less reliable and more dangerous member of the community. In important matters, the demands of conscience must always be supported by legal sanction.

20. In the United States, population problems are problems of preserving amenities, rather than mere necessities. The fact that people are not starving here (even if true) does not mean we do not have a population problem. In a rich country, increased population means decreased privacy, decreased freedom, a decreased share of the nondivisible goods of the world, and an increased "information overload."

21. The good life must include a share of nondivisible goods. Solitude by definition cannot be shared. The qualities of wilderness and the outdoors, of canoeing and beaches and nature hikes are lost with crowding at a very low level. With a population the size of ours (or even a small fraction of it) a fair division would destroy the goods themselves. We cannot build a four-lane highway into the wilderness.

22. For nondivisible goods to be enjoyed by any, they must be restricted to only a few. Basically there are only two methods of restriction: (a) by lottery and (b) by merit. Consider the goods we call wilderness. We can allow entrance into it on the basis of a lottery. This will be regarded as fair, but it will have the disadvantage that it will admit many people into the wilderness

who are not prepared to take maximum advantage of it. Bluntly put, some will be physically unfit to enter a primitive area. It would be more sensible to admit on the basis of physical merit—that is, physical fitness. Various ways might be used to determine this. Objective tests might be administered. Or we might simply make it physically difficult even to get to the wilderness, e.g., by stopping all roads ten to twenty miles short of the wilderness area, and then making people walk the rest of the way. The idea of privilege based on merit goes counter to dominant trends of the last fifty years. But trends can be reversed.

23. When we succeed in reaching an acceptable definition of the optimum population, there is little doubt that we will find that we have already overshot the optimum to a considerable extent. After achieving zero population growth, we will have to achieve a negative rate of growth for a long while. The cost and pain of a negative rate of growth will no doubt be considerable. This is a problem that we someday must tackle.

24. Every "need" has at least one contradictory alternative that should be considered before reaching a decision. It has been said that we "need" more oil; it can just as logically be said that we need a smaller population, fewer autos or more efficient transportation. "Need" is a treacherous word.

25. Just as mass starvation cannot be eliminated by food, so also traffic problems cannot be eliminated by building more roads. More roads merely increase the amount of traffic and re-establish the problems at a higher level, after diminishing the amount of open space available for human enjoyment.

26. In the early stages of population control it may, for political reasons, be necessary to use largely indirect methods. Among these might be refusal to grant building permits, making greater use of restrictive zoning, refusing to build more roads, and establishing exorbitant taxes on gasoline, thus encouraging people to live nearer their work or use mass transportation.

27. Among the indirect controls of population growth that we should work on immediately is a modification of education. At the present time, from kindergarten through high school, we hold up only one ideal of the good life: the ideal of having a family. But not everyone is psychologically fitted for family life, nor is there any community reason for wanting to force everyone into this mold. Unfortunately, all public school education implicitly

presents a single ideal to a richly variable population, and thus forces many essentially non-parental types into the parental mold. We need to tell little children that it is also possible to live a good life without having children. We should not displace an old ideal, but augment it with additional possibilities.

28. Ultimately we may well have to come to direct coercive control of individual breeding. Battles will no doubt be fought over this proposal. Looking forward to this time, the disciplines of political science, psychology, and biology should all be combined in a creative effort to discover those social inventions that will be least painful and most acceptable in achieving direct control of population.

Fourth Group

Is that all? Certainly not. The catalogue of heresies just given will no doubt seem shocking enough to people who are involved in public education. Nevertheless, there are more. The catalogue is open-ended.

We should seek at all times to tell "the truth . . . and nothing but the truth"—but we are under no compulsion to tell "the whole truth" *immediately*. Beyond the heresies given here lie other heresies far more shocking, heresies for which the world is not prepared. I shall not even mention them in public.

There are still other truths that are too shocking even to be mentioned in private. If I may be permitted to make a single sibylline utterance let me say just this: "The ratio of two exponential functions is itself an exponential function." The social implications of this mathematical truth in a world that is finite, are radical in the extreme, and cannot be so much as mentioned without violating cherished taboos of our time.

So I had better stop.

CHAPTER FIVE ✳POPULATION ✳ *Comments on The Hardin Paper* by HAVEN KOLB

The sweep of Dr. Hardin's theme, the multiplicity of ideas that he has showered upon us in a compact presentation, combined with the novelty of some of them, may have—for many readers—an overwhelming effect. There are a few points on which comment may be useful.

The initial statement requires some qualification. Not long ago in writing a paper for the silver anniversary publication of the National Science Teachers Association I had to think through the matter of pressures on the educational process. My first position was much like Dr. Hardin's, but eventually I came to consider it oversimplification. A society does not deliberately finance an educational system that is dedicated to its own destruction. But I may support a system that leads, consciously or accidentally, to the evolution of the present society into a different one. I think the terms "finance" and "educational system" must be interpreted broadly. If so, the Hopi may be said to have financed an educational system that so firmly undergirded the establishment that three centuries of pressures from two different European cultures have made little impression on it. However, what Ruth Benedict called the Dionysian worldview of the plains tribes allowed Pawnee and Comanche to seize upon only one aspect of the European advent and by rapid acquisition of equestrian skills effect a radical shift in their cultures. Space restrictions require that this brief historical allusion serve in place of argument.

There is a further difficulty with the initial statement. It seems to close the door on the possibility that a teacher, especially one supported directly by the public, can do anything to change the status quo. Yet the major part of the paper advocates a course that condemns the cherished beliefs of the teacher's paymaster. We get an acknowledgement that this presents a dilemma, but the dilemma is then brushed aside by a figure of speech. Leaving out of consideration any moral aspects of the dilemma, we need, I think, considerably more attention to the practical aspects of staying employed while promulgating the heresies.

Now concerning these heresies, is it significant that they have been so labeled? A heresy is merely an opinion held in opposition to commonly received doctrine. Neither "doctrine" nor "opinion" have any necessary connection with truth. If the goal is good perhaps we need not worry too much about the percentage of objective truth in our heresies. Nevertheless, I am bothered. I wish some were stated in less polemical form and I have definite reservations about Number 2. In Europe, a large upswing in population began considerably before the time of Louis Pasteur.

And I must say a disquieting word about another of the heresies. That "space" is no escape from our population problem may be a completely convincing proposition for Dr. Hardin—and it is for me. But it is not at all convincing to a great many voting citizens. And they are encouraged in their vain optimism by the pronouncements of many prestigious scientists. Some heavy fighting is due in the main arena before we can push this one very far in that least credible of contemporary sources of information, the classroom.

There are, of course, numerous sectors along which advancing biological knowledge subverts the social structure. But the impact of the population problem seems to me to force all other social issues into subsidiary positions. Certainly there is no reason to believe that the stabilization of the global human population would solve all social problems—war, for example, seems to have existed even where populations have been extremely sparse. But without population stabilization—and that very soon—it seems clear that attendant social problems such as pollution, the attenuation of resources, and the acerbation of perennial nutritional deficiency will leave room neither for attack upon older social problems nor for more than perfunctory attention to the implications of other biological matters such as radical surgical techniques and genetic engineering. Some of these last-mentioned matters have captured a great deal of attention, but let us put first things first.

CHAPTER FIVE ✶ POPULATION ✶ *One More Key Truth to Teach About Population* by LAWRENCE MANN

In general, I think Dr. Hardin's approach should be very valuable to teachers who would like to dare tell more about population to their students. His rhetoric is nicely arranged to build up courage, and he has set down most of the hard truths that need to be told.

Of course, many of the things Dr. Hardin writes are not at all so shocking as he seems to think they are. Particularly, some of the points he considers most heretical are surprisingly tame. Yet it is not in these points of fact and their organization that I must take issue with Dr. Hardin.

The main problem with his paper lies in one major misconception and one related sin of omission. So crucial are these that, taken together, they make his approach defectively oversimplified.

Dr. Hardin sets his stage by evoking the specter of a pervasive establishment that tolerates scholarly search for truth up to some point of heresy, beyond which scholarship will be stamped out. The teacher of population biology is invited to participate in this delicious danger by working from a "graduated checklist of heresies."

The implication in this little foreplay is that the establishment does not want the hard truths about population to be told and that it actively opposes population control. I fear that these assumptions amount to a very serious misconception. A generation ago they would have held, even a decade ago it might have been hard to question them. But as the 1970's begin in this country and the world, indications are that whatever establishment there may be is increasingly in favor of population control—and the use of shocking truth to foster it. If population teachers are endangered by telling the truth, the danger will come from concrete human groups rather than abstract establishments.

For simplicity, and to carify the ultimate perimeters of the population problem, Dr. Hardin takes an essentially one-world view of the human population. But an important complication of the human population problem is that humanity has subdivided itself into races, ethnic groups, nations, etc. Some of these have more economic resources and political power than others; and, through much of human history, the advantaged groups also had faster population growth rates (net) resulting in a neat preservation of a *status quo*.

What preventive medicine on a pervasive scale has done is to increase the net surviving population most rapidly in those races, ethnic groups, and nations that earlier were held in check by disease. (Advantaged groups were less affected because of better nutrition.) The bulk of the human population explosion is taking place among such disadvantaged groups, and such groups doubtless feel that increasing numbers will bring increasing strength. The bio-ecologist knows that when two groups compete for the same resources, the group with the higher growth rate will dominate unless one starts eating the other. The same idea has been current among Caucasian peoples for some decades now; e.g., "the yellow peril."

National and international establishments are certainly involved in promoting population control, especially among disadvantaged groups. What has happened in U.S. foreign policy in this regard during the past decade is quite clear. Gentle but firm pressures are being applied to almost all countries receiving foreign aid. This policy appears to be having very limited impact on population growth rates, but that is probably because economic missions and embassies, politically, dare apply no more pressure. Similarly, the increased attention to population control in relation to black and other minority recipients of welfare funds is very much in evidence in the United States. There have been more than a few cases in which black women were actually denied welfare for failure to attend a birth control clinic.

The timing of interest in population control by U.S. Government agencies cannot escape the notice of disadvantaged peoples in developing countries and in urban racial ghettoes at home. Overseas, it came after it was clear that Alliance for Progress and similar schemes were not going to bring the dramatic results in developing countries that the Marshall Plan had in Europe. In the United States, interest in population control for disadvantaged minorities came after the failure of the War on Poverty to show signs of dramatic accomplishment.

There have been consequent reactions to the population control efforts of U.S. agencies and the American foundations. These reactions are more muted abroad than at home. Black militants are increasingly taking a much stronger stance against population control than that of any of the religious or ideological opponents of it. The teller of population truths can easily find himself labeled a racist advocate of genocide against black people, with suitable semantic embellishments.

I insist that the establishment is apparently more and more in favor of population control, including the use of Dr. Hardin's "heresies" as propaganda. The main truth I want to add to Dr. Hardin's checklist is that population control can be used to maintain a politico-economic *status quo* among advantaged and disadvantaged groups of a human population. How efficient it may be in this respect is not as important as the fact that it is viewed as having this potential effect by both groups.

The long-range dimensions of the human population problem remain. Dr. Hardin's strategy is probably a good one for teaching the fearful facts on this subject. But the short-range complexity of the politics of population propaganda and population control alluded to above must also be included in any checklist of hard truths. If the long-range view is to have the necessary impact (and I am more pessimistic than Dr. Hardin on the inevitability of this) it must be argued in the context of short-range political truths.

CHAPTER FIVE ✳ POPULATION ✳ *The Need for Ecology and Social Biology in Liberal Arts* by CHARLES H. SOUTHWICK

Although I'm sure Dr. Hardin and I agree on most fundamental aspects of the population problem, for the purpose of discussion I disagree with some of his current approaches.

Let us first consider some of our points of agreement. It is obvious to both of us that the world population is drastically out of balance, and is headed on a collision course with the facts of ecology. The population of the world is increasing at the rate of more than one million people per week, yet we cannot adequately provide for those already on earth. More than 50 per cent of the world's population is underfed, poorly housed, ill-clad, improperly educated or not educated at all. Medical services, social services, recreational opportunities, and freedom of expression are more drastically curtailed for many of the world's peoples now than ten years ago. We are caught in a deadly vise of increasing population and deteriorating environment, but the economic bulldozer of population expansion continues to push us toward a future with even fewer options than we now have. On all these points, I think we agree. Most ecologists have, in fact, collectively worried about this for twenty years, and we're tired of hearing ourselves go over these same old problems.

As a matter of fact, however, I don't think these points of view are generally accepted, and I think even Group One of Dr.

Hardin's heresies are far from common knowledge. We still have economists such as Colin Clark of Oxford telling us that there is no such thing as a population growth problem; that the world can easily support 40 billion people; and that rapid population growth is essential for economic growth. We still have the Chiefs of State in some Latin American countries saying that the basic problem in their countries is insufficient population growth. Most scientists feel that quite the reverse is true. No wonder the issues are clouded, and no wonder the public is confused!

Dr. Hardin's approach is now the shock approach—even the cryptic shock approach, for did he not write that there are other heresies far more shocking, which he cannot even mention in public or private! What are we to make out of this? Is he finally becoming facetious about the whole matter? Or is he implying that he understands the solutions but they are too sophisticated for us? This approach will clearly make more enemies than friends. I think there is considerable evidence that the shock approach does not work, not even with the public and certainly not with scientists.

I think we are still within the range of some solutions to man's ecologic crises, though we may not be for much longer. One solution that I feel is necessary, and we should in fact have started thirty years ago, is to begin teaching more ecology and social biology at all educational levels from elementary school to adult education, and subsequently to structure all of our educational efforts around the theme of man and his environment. I include man's physical, biologic and social environment. History, economics, social studies, humanities, and the natural sciences should all be built around the theme of ecology and social biology. As such ecology and social biology become core subjects in a liberal arts education. I consider it a sad and untenable position that most of our undergraduate colleges and universities still do not have courses in ecology, and still do not link ecology and sociology in a proper way

As Aldo Leopold said twenty years ago, we must build into our citizenry an ethical concern for the landscape, an ecological conscience, so that we recognize mankind as a member of the earth's biotic community and not its sole owner and master. With this educational approach, we will be training business and political leaders, lawyers, engineers, teachers, ministers, doctors, clerical workers, and factory workers the facts of life, so to speak, by which I mean the facts of ecologic life. We must stop extolling *ad nauseum* the great and wonderful accomplishments of man's technology. As marvelous as these are, they are misleading unless the other side of the coin is also presented with equal force, unless we are made to realize that man still depends upon the thin blanket of the biosphere, a blanket that is becoming more shredded and torn every day. Only then can the public make some of the hard political decisions mentioned by Dr. Hardin which are so drastically needed.

I agree that we must bring population growth down to zero; that we must cease the destruction and pollution of our environment; and that we must achieve ecologic and social equilibria that will permit the continuation of life on earth. But if all of these things are to come to pass without an absolutely totalitarian state, they must come through a sensitive, responsive, and very intelligent populace. We need more ecology and social biology in our curricula at all educational levels to create this populace.

CHAPTER FIVE ✷ *Discussion*

QUESTION—*Is anything being done about reducing population growth?*

DR. HARDIN: There has been formed a group called Zero Population Growth, Inc. which is tackling the intellectual question, what are the problems of having a zero population growth rate? The group has initiated a congress in Chicago June 7 to 11, 1970, to deal with this problem. I've tried for ten years to get my economist friends to take seriously the problem of zero population growth and study seriously what is the optimum population. They've told me that this was a meaningless question and could not possibly be attacked. I don't believe it, and at last some of the economists—not many—are beginning to see that this is a meaningful question and should be attacked. This is on the intellectual level. What you do on the action level, since we are just beginning, is to flounder around a bit and discover a few things. I think what is most needed is that each person should have something of an ecological conscience about what goes on in his community. In my community we're currently having a battle over whether a four-lane freeway shall be built around the University and destroy a slough.

On the West Coast sloughs and estuaries are very scarce. There are only four sloughs between San Diego and San Francisco and ours is one of them. The biologists prize it, and now we are faced with a plan to destroy the slough with a road to take people past the University at sixty miles per hour. The biologists regard this as utter insanity and we're fighting it. The Biology Department and two thousand of our thirteen thousand students have signed a petition against it. Unfortunately, our Chancellor (who, I am sorry to say, is a biologist) is a growthmanship man and believes in progress with a capital P. He thinks anything that involves an expenditure of 6 million dollars is necessarily wonderful. We had better look at our local battles and estimate whether we have a chance to win them. We should fight for small gains on the local scene. It is not possible for one person to save the world; each should do his bit.

QUESTION—*One of the points mentioned was population control by coercion, e.g. a population tax. But wouldn't this be primarily a burden on the poor?*

DR. HARDIN: Are you saying that the rich are proposing to coerce the poor into having fewer children for the sake of the rich?

QUESTION—*No, that is not what I mean. Assume the government is going to enact legislation to control population. If the government is going to coerce people not to have children, and if a population tax seems to be the easiest way to do it, then it would appear to be discriminatory against the poor. It seems to me we are going to have to rectify other social problems before we can attack the population matter in this way. Now I don't think that this is something being deliberately proposed by the rich. I don't think the rich want to have that many more children; they're simply trying to keep the poor from breeding too much. I think we are not looking at the problem closely enough.*

DR. HARDIN: I imagine that what we do should probably take place in stages. The first stage, I would guess, would be that we would finally, after considerable trial and tribulation, arrive at the idea that every family is entitled to have say, two children, if it wants to. Two children "on the house," so to speak, but no more; and there would be no distinction among families. The rich can't have more just because they have more money. If they have more money they also occupy more of the environment. Their cars go faster; they have more cars; they produce more pollution. That is considered "not fair." If you're rich you have two children; if you are poor you have two children, the same for everybody. My guess is that this would be the first stage.

We may later decide that some people, maybe, ought to have a few more. We might discover a Beethoven and decide it would be nice to have another Beethoven. Of course, we'd get into battles because some people would want Beatles instead of Beethovens.

Incidentally, as far as the poor are concerned, many people think that the way to get them to have fewer children, so that they are not as large a burden on the tax rolls, is to distribute contraceptives widely. However, the problem is far more complex than this. The desired goal of a reduction in their bearing of children would be achieved much better by seeing to it that every poor woman has alternative goals. If a poor woman has the real opportunity to work in a factory where she can socialize with other women at coffee breaks and lunch time; if she can thus increase her family income; and if she can leave her children at child care centers while she works—then I think you will find that most poor women given that alternative to staying home and raising a large family will choose the factory work. It is much more pleasant.

DR. SOUTHWICK: None of us can predict the level at which the human population will reach an asymptote or the means by which it will be reached. Raymond Pearl tried such an estimate back in the twenties and he predicted that by the year 2000 the world population would be 2.5 billion, so his prediction will be incorrect by several billion people. Obviously the population is going to be controlled by some means at some point in the future and the question at hand, I think, is how much human tragedy must we go through and how much misery will we permit or tolerate before population control is achieved either by drastic catastrophic means —starvation, disease, social collapse, massive war, etc.—or by rational means.

I would like to raise one further point and it relates to the question of activism. I think our greatest role and our greatest influence is going to be as teachers. If we can teach some of the principles of population ecology, and if we can properly draw the lines of connection between population and environmental deterioration, pollution, and social degradation, as these are properly related, we can have a fantastic influence on the thinking of this country. The NABT has ten thousand members and each of us, as teachers, has an opportunity to influence at least one hundred people a year, often many more. That amounts to one million people a year. Thus this organization can exert a tremendous scientific and social influence as individual teachers.

QUESTION—*What information is currently available about the effects of population pressure on human beings?*

DR. MANN: The problem with what is known or what is said about the effects of population density, or intensity of interaction of human beings, on human behavior is that there is imperfect agreement. The information available is very vague and suspect and is bandied around among scientists to such a degree that I don't think there is any satisfactory material that could be offered at this time for teaching in the high schools. The anthropologist, Edward T. Hall, summarizes some material, much of it borrowed from animal behavior studies, in *Hidden Dimensions.*

DR. SOUTHWICK: I heartily agree with Dr. Mann that we just don't have enough facts but there are some remarkable correlations. There are studies which have received little distribution and publicity so far, demonstrating elevations in chronic disease, premature hypertension, and rates of mental illness in inner city environments. We recognize, on a world-wide basis, an inner-city syndrome which involves high rates for disease, prenatal loss, mental illness, crime, drug abuse, social disruption, and family break-ups. From a research point of view the problem is how to dissect and isolate the influence of crowding from other influences such as environmental deterioration, malnutrition, and substandard educational opportunites. There is enough now known to make an effective teaching story if the information were more widely distributed from the medical journals into more generalized biological literature.

QUESTION—*Why did you say that birth control is not the equivalent of population control?*

DR. HARDIN: This point is one that I borrowed from Kingsley Davis, who has made it very well. The answer is quite simple. We have appropriate data now from many countries on two statistics. First, you ask people what they regard as the ideal size of a family or, to put it in another form, how many children do they intend to have in their family? (These give slightly different answers but it doesn't matter which you take.) Now that is one statistic. The second statistic is this: in the light of present health conditions in that country, what number of children is needed to maintain the population at zero population growth rate? For literally dozens of countries now we have these two statistics and, without exception, the number of children people feel is ideal or the number they intend to have is always larger than the number that would keep the population at a constant level. This is a simple way of showing that family birth control is not national population control. People who have satisfactory birth control are able to have as many children as they want and that is too many. The moral of this is: to achieve population control you must not only have birth control but you also must have some way to alter people's ideas of what is proper. Their ideals must be changed or population control will be impossible.

CHAPTER SIX ✻ EVOLUTION

CHAPTER SIX ✻ EVOLUTION ✻ *Evolution: The Reluctant Revolution* by CLAUDE WELCH

A few years ago the Bishop of Woolwich, Dr. John A. Robinson, wrote a little book. The book was titled, *Honest to God.* And it was a very honest book indeed. Dr. Robinson tells it like it is as far as he is concerned.

The reaction to this little book might well be a song titled "Oh Dr. Robinson."

In fact, some of his parishioners and fellow clergy undoubtedly felt as though Dr. Robinson and the famous Mrs. Robinson of "The Graduate" fame would make a good pair. Dr. Robinson has probably been paired with the Devil, International Communism and Ho Chi Minh depending upon the pet peeve of the particular critic.

What does Dr. Robinson have to say about evolution and its social implications? Practically nothing in particular, but practically everything in general. He speaks of a reluctant revolution which is going on in theology. We will speak of the reluctant revolution of evolution. He speaks of asking honest questions; questions which almost everyone poses to himself, but rarely expresses in public. We will ask some honest questions about the interaction of science and society and the relationship between theoreticians and theologians.

Dr. Robinson asks some honest questions about the faith and we will ask some honest questions about evolution theory. He doesn't call his faith a theory and we won't call our theory a faith. But we will explore the relationship between faith and theory and thus will ask a few of the perennial questions about science and religion.

Evolution has two kinds of social implications. One kind involves the impact of the idea itself, both on and within the scientific society and on and in reaction with the non-scientific society and its culture. The second kind of impact extends beyond the idea itself and involves this question: If evolution has occurred and continues to occur, what might be its possible consequences on the future history of man? We can have similar concerns about an organism: what goes on within the organism; what goes on between the organism and its environment; and what will go on during its historical development? It is certainly an open question as to which has had the greater implications since 1859: evolution itself, or the idea of evolution. As any child will tell you: if there is anything worse than a hypodermic needle, it is the idea of a hypodermic needle.

So this will be our first task: to explore why evolution has been so reluctantly accepted by society in general and a portion of the scientific society in particular. This will involve a discussion of the nature of science and its ideas. Our second exploration will involve a study of the future implications of evolution theory if one accepts its postulates as feasible and the evidence as sufficient.

There are some, no doubt, who will argue that there are no relations between science and religion and that the two disciplines will continue to go their separate ways. But others have said otherwise. Alfred North Whitehead in his book, "Science and the Modern World," says: "When we consider what religion is for mankind and what science is, it is no exaggeration to say that the future course of history depends upon the decision of this generation as to the relations between them." Now it is the tactic of the lazy argument or *ignava ratio* to say simply that science and religion will go their separate ways. And this may well be true. But if Whitehead is correct in his estimate of the importance of the relations between them, a lazy argument will not do the trick. Now we may end up after our analysis just as lost as those who refuse an analysis. But that is the chance you take and you accept the small profit in knowing *why* you are lost instead of knowing simply that you *are* lost. And besides, as suggested by Robert L. Stevenson, "To travel hopefully is a better thing than to arrive."

Reluctant revolutions seem hard to come by. This is because we usually think of revolutions in terms of brute force, violence, and strife. But conceptual revolutions are a different breed of cat.

They are often very subtle in that they proceed in small quantum jumps—one man at a time. A conceptual revolution is infectious, but rarely pernicious. The infection spreads and raises a fever in those who are particularly suspectible, but often lies dormant as in lysogeny, coming to fruition when the time is ripe. Rarely does Saul become Paul in a flash of insight.

We shall first take a look at the impact of new ideas on science itself. This may give us a clue as to the reason for the delayed acceptance of evolution by the general public. It may also give us a little deeper insight into the nature of science itself. And we may even consider that in the teaching of evolution in our classrooms, we may find that a more intensive look into the reasons for the acceptance or rejection of a thoery not only helps us to understand evolution, but also helps us to understand the nature of science and its relation to society. It seems to me that a creative teacher might tell us that when faced with a controversial idea one should use it to good advantage. If the idea is controversial it will hold the student's attention and interest and an additional, or perhaps initial, contact is made. If you have the student's interest then you can explore the nature of the controversy and the nature of the adversaries. In the long run you will not only learn a lot about evolution theory (what it is and what it isn't) but also a lot about the nature of science, as well as the nature of the ideas held by society in general and by students in particular. In short, I have seen the goals of a biology course stated as follows: To understand the nature of biology and its relations to the other sciences, and the nature of the relation of these to society. It would be difficult to find a more inclusive vehicle than evolution theory for the attainment of these goals.

Thomas S. Kuhn, in his book *The Structure of Scientific Revolutions,* contrasts the scientist as he often sees himself with the scientist as he appears to others. The scientist often expresses his warmest regard for the honest search for truth wherever this may lead him. He pictures himself as the knight in shining armor ever ready to slay the dragons of ignorance and superstition. He is ever ready to throw off old ideas and adopt new ones if they are better able to account for phenomena. The sad truth is that the history of science shows us that most scientists are dreadfully unwilling to shake off a comfortable theory even with a knowledge of its incompleteness and its inability to compete with a newer idea. Objectivity in science is not necessarily a myth, but it is often quite elusive. We should not be surprised for, as James Conant has suggested, scientists seem to cover the whole spectrum of human folly. An objective analysis of our friends and acquaintances would certainly support Conant's view.

In most disciplines the young scholar appears on the scene as a giant killer. With tireless endurance and religious zeal he enters the battle to do service to the cause of truth. He questions the basis of just about everything and is often particularly eager to challenge the giant ideas which are the footings of his discipline. But he soon tires and becomes absorbed in what is known as normal science. Kuhn puts it this way: "Normal science, the activity in which most scientists inevitably spend almost all their time, is predicated on the assumption that the scientific community knows what the world is like. Much of the success of the enterprise derives from the community's willingness to defend that assumption, if necessary at considerable cost. Normal science, for example, often suppresses fundamental novelties because they are necessarily subversive of its basic commitments."

Kuhn's statement reminds us of a burlesque attributed to Francis Bacon.

"In the year of our Lord 1432 there arose a grievous quarrel among the brethren over the number of teeth in the mouth of a horse. For thirteen days the disputation raged without ceasing. All of the ancient books and chronicles were fetched out, and wonderous, ponderous erudition such as was never before heard of in this region was made manifest. At the beginning of the fourteenth day a youthful friar of goodly bearing asked his learned superiors for permission to add a word, and straightway, to the wonderment of the disputants, whose deep wisdom he sore vexed, he beseeched them to unbend in a manner coarse and unheard-of and to look in the open mouth of a horse and find answer to their questionings. At this, their dignity being grievously hurt, they waxed him exceeding wroth: And, joining in a mighty uproar, they flew upon him and smote him, hip and thigh, and cast him out forthwith. For, said they, surely Satan hath tempted this bold neophyte to declare unholy and unheard-of ways of finding truth, contrary to all the teachings of the fathers. After many days more of grievous strife, the dove of peace sat on the assembly, and they as one man declaring the problem to be an ever-lasting mystery because of a grievous dearth of historical and theological evidence thereof, so ordered the same writ down."

This burlesque was made obviously as a plea for experimental science as opposed to dogmatic conceptual systems. But the paragraph can have a deeper meaning when applied to the intellectual inertia which often besets an established discipline. "If you have had your attention directed to the novelties of thought in your lifetime," says Whitehead, "you will have observed that almost all really new ideas have a certain aspect of foolishness when they are first produced, and almost any idea which jogs you out of your current abstraction may be better than nothing. We need to entertain every prospect of novelty, every chance that could result in new combinations. But at the same time, we need to entertain those with skeptical examination and subject them to the most impartial scrutiny, for the probability is that 99 per cent of them will come to nothing, either because they are worthless in themselves or because we shall not know how to elicit their value; but we had better entertain them all, however sceptically, for the next idea may be one that will change the world."

The editors of the *Physical Review* hope they have alleviated this problem because they claim that most of the crackpot papers which are submitted are rejected, not because it is impossible to understand them, but because it is possible. Those which are impossible to understand are usually published. How often do we find a conversation similar to Niels Bohr's remark to Wolfgang Pauli: "We are agreed that your theory is crazy. The question which divides us is whether it is crazy enough to have a chance of being correct. My own feeling is that it is not crazy enough." We find Lorentz, the Dutch physicist, reminding us, "One of the lessons which the history of science teaches us is surely this, that we must not too soon be satisfied with what we have achieved. The way of scientific progress is not a straight one which we can steadfastly pursue. We are continually seeking our course, now trying one path and then another, many times groping in the dark, and sometimes even retracing our steps. So it may happen that ideas, which we thought could be abandoned once and for all have again to be taken up and come to new life."

A look at a few of the scientific revolutions, like evolution theory, makes us realize that an important scientific innovation rarely makes its way by gradually winning over and converting its opponents. It rarely happens "that Saul becomes Paul. What does happen is that the opponents to a new idea gradually die off and the growing generation is familiarized with the idea from the beginning."

Now this point has important implications, both in science and in society, and also as to how the two relate to each other. Spontaneous generation was a perfectly acceptable idea in an earlier century—acceptable by both science and society, because it fitted the belief-expectancy patterns of both. In a later century the idea was gradually rejected by both groups, but not without great, long controversies. The notion has now been reintroduced and has come to new life. If a student can see how ideas, like spontaneous generation, evolution and natural selection, are products of the human mind and not facts in and unto themselves, he will perhaps begin to see science in a new light, a light that reveals its humanistic dimensions, dimensions with which the student can partially identify.

Julian Huxley tells the story that when he and H. G. Wells were writing a book together, Huxley had calculated if all of the red blood cells in a human body were placed next to each other in a single line, the line would stretch several times around the earth. Wells said, "Oh, we couldn't put that in. The students would never believe it and then they would never believe anything else in the book."

Now this Wellsian attitude often prevails concerning the nature of science. Don't tell the kids that some practitioners of science change their minds; that some refuse to change their minds even in the face of overwhelming support for an opposite position; and that some cheat. Shouldn't students understand that scientists are mortal—and fallible—human beings?

I'm sure I'm belaboring the point. I am trying to say that the revolutionary idea of evolution, as well as many other revolutionary ideas, have come painfully slowly into the sicentific world for all kinds of reasons, some scientific, some emotional, some childish. I am saying that students can learn much about the nature of evolution and the nature of science by being informed about this aspect of science. Furthermore, it may help us to be a bit more charitable to the general public (another group of human beings) and thus to at least understand why some people are reluctant to accept the idea of evolution. If scientists can be

stubborn when they have a substantial array of facts at hand, can
we not understand a reluctant public with less information at its
disposal— especially when confronted with an idea which is simple
in its essence but complicated in its application? Should we be
surprised at the reluctant acceptance of new ideas by laymen when
members of the scientific community seem no more flexible in
this respect? Kuhn again remarks: "The more carefully the
historians study, say, Aristotelian Dynamics, Phlogistic Chemistry,
or Caloric Thermodynamics, the more certain they feel that those
once current views of nature were, as a whole, neither less
scientific nor more the product of human idiosyncrasy than those
current today. If these out-of-date beliefs are to be called myths,
then myths can be produced by the same sorts of methods and
held for the same sorts of reasons that now lead to scientific
knowledge. If, on the other hand, they are called science, then
science has included bodies of belief quite incompatible with the
ones we hold today. Given these alternatives, the historian must
choose the latter. Out-of-date theories are not in principle unscientific because they have been discarded."

"Copernicus, Newton, Lavoisier, Einstein [and most certainly
Darwin] —each of them necessitated the community's rejection of
one time-honored scientific theory in favor of another
incompatible with it. Each produced a consequent shift in the
problems available for scientific scrutiny and in the standards by
which the profession determined what should count as an admis-
sible problem or as a legitimate problem-solution. And each
transformed the scientific imagination in ways that can only be
described as a transformation of the world within which scientific
work was done."

How, then, are scientists brought to make a change in their
world view? Part of the answer is that they are very often not.
Copernicanism made few converts for almost a century after
Copernicus' death. Newton's work was not generally accepted,
particularly on the continent, for more than half a century after
the *Principia* appeared. Priestly never accepted the oxygen theory.
The difficulties of conversion have often been noted by scientists
themselves. Darwin, in a particularly perceptive passage at the end
of his "Origin of Species," wrote: "Although I am fully convinced
of the truth of the views given in this volume . . . , I by no means
expect to convince experienced naturalists whose minds are

stocked with a multitude of facts all viewed, during a long course of years, from a point of view directly opposite to mine . . . . But I look with confidence to the future—to young and rising naturalists, who will be able to view both sides of the question with impartiality."

Darwin's theory of evolution by natural selection is one of the three major paradigms in biology. A paradigm is a conceptual scheme or theory which attracts an enduring group of adherents and is sufficiently open-ended to leave all sorts of problems and puzzles to be resolved by the practitioners of normal science. It is clear that the evolution-natural selection paradigm was adopted by the impartial judgment of the young and rising naturalists referred to by Darwin.

The reason for the selection and rejection by scientists of scientific theories or paradigms is certainly not unrelated to the subject at hand. We have suggested that the transition from one system of beliefs to a new one is a long and difficult process. The famous Ptolemaic-Copernican controversy is one of the most complete case histories available to us. It was a reluctant revolution for the very reasons we have been discussing. No theory ever solves all the puzzles with which it is confronted at a given time; nor are the solutions achieved often perfect. Of one thing we are pretty certain: competition between segments of the scientific community is the only historical process that ever actually results in the rejection of one previously accepted theory or in the adoption of another. First, a theory is not dropped simply because it can't account satisfactorily for all of the known facts. Second, it is difficult enough to replace an old theory with a new and better theory, but this can be done given enough time. Third, a new theory usually develops (during a crisis period) when an old theory cannot account for novel observations arising through the activities of normal science. But the new theory must be a theory in the full sense of the word and thus subject to the same treatment as the old theory. This last series of statements accounts for the almost universal misunderstanding of the relation between science and religion in general, and biology and the Judeo-Christian dogma in particular.

It cannot be emphasized too strongly that competing ideas in science involve theories and their goodness-of-fit to phenomena and observations. There are certainly other elements involved in the final adoptions, but the rules are fairly clear-cut and the battle proceeds through experiment, counter-experiment, charge and counter-charge. It is clear to the proponents involved that each theory consists of a series of assumptions or postulates, and although the postulates are not capricious, they are certainly subject to change—almost without notice. As the theories compete for acceptance, as new experiments are performed, as new facts are uncovered, the postulates are continually refined by change, addition, and deletion in order to provide the simplest, yet most inclusive assumptions possible. Sometimes the theories even tend to merge when it is found that a compromise set of postualtes seem to be most representative of current observations.

Now contrast this rather pragmatic approach to reality with a system involving dogmatic theology. A poet once told me that the difference between a scientist and an artist is that a scientist searches for the truth whereas the artist knows the truth and his concern is to tell us about it. How can there be a competition between a scientific theory and a dogmatic theology when the two belong to two different epistemologies? Scientists freely admit that their theories contain a body of postualtes which are basic assumptions. These postulates are a product of the human mind, subject to change, and have no extra-terrestrial source. A religious system, on the other hand, consists of a series of affirmations, rarely subject to change, and claims for divine revelations are made for them. It is clear that this is not a fair debate because there are no rules for such a debate. The theory is subject to change but committed to an attempt to explain phenomena. The religious system is not subject to change and is only partially committed to explaining the world of objects and things.

So I agree, in part, with the lazy argument approach. Scientific theories and religious systems will go their separate ways because they have a minimal contact zone. A theory can only compete with another theory and, I suppose, a religion can only compete with another religion. Now when a scientific theory or science itself, becomes a religion, called scientism, then the social implications are clear. The postulates become affirmations, the voices of the scientists become strident and prophetic in tenor and theories become facts. One can identify fundamentalists in both science and religion.

When people talk about the conflict between science and religion, they almost always mean between biology and the Judeo-Christian religion, specifically, genesis. So it is not fair to implicate all of science and all of religion. Physics, chemistry, and astronomy have little conflict with the religions of today, and many of the more other-worldly religions have little contact with scientific ideas. Even when biology and the Judeo-Christian religion are contrasted, the key issue usually involves evolution and genesis. Now a solution of the issue could be approached if both evolution and genesis could be considered as scientific theories. Let us consider both as a series of assumptions and then proceed to correlate them with known facts. (There are a few individuals who feel that if evolution theory is presented in the public schools, then the genesis account in the Bible should be given equal time. Again, there is no point of contact between a scientific theory and religious dogma.) But if a group of scientists care to propose a genesis theory in competition with evolution-natural selection theory, then it is fair game. A group called the Creation Research Society does publish a journal in which a realignment of science is proposed based on theistic creation. The Society affirms "To the student of nature this means that the account of origins in Genesis is a factual presentation of simple historical truths."

It is my impression that this group would not be willing to propose a genesis theory that could be contrasted, in a competitive sense, with an evolution-natural selection theory.

Yet it is also my impression that theoreticians and theologians have much in common. They all seem like intelligent people trying to make some sense out of the universe. The theologians move closer and closer to viewing their religions as postulational systems and to the position that their affirmations might be open to reinterpretation. Certainly Mr. Robinson, in his *Honest to God* book, gives one the impression that the reluctant revolution in theology will propose some drastic changes in viewpoint. Am I saying that science is right and religion is wrong? No. Judging from what I presently know of scientific concepts, those we have now, if not wrong, at least are only partially right. Judging from what I read from the theologians, many are convinced that a reorientation of some very basic religious ideas are in order. And, just as in science, the revolution will be, and is, reluctant and sometimes painful. There will be even louder cries to defend the faith. When our objectives become clouded we usually double our efforts to hold onto them.

In proposing the importance of a decision by this generation as to relations between science and religion, Whitehead was emphasizing the implications of competing philosophical systems. Science has always been a humanistic enterprise and our religious systems are certainly moving in that direction. What difference does it make to man, and to society, as to the relations between these systems?

Evolution is a revolutionary idea of really awesome dimensions. The Copernican Revolution, which took man out of the center of the universe and put him on a small planet near a third class star located at the edge of a minor galaxy, was a major turning point in man's intellectual history.

To the evolutionist, man is not the center of a paradise created specifically for him. He comes late to this earth and has imposed himself as a masterful predator. His history is essentially the history of warfare. Reinhold Niebuhr has suggested that if we believe man's historical existence to be meaningful, we do so by faith and not by reason. It seems to us that evolution is not predestined to produce always the good and the beautiful. The Darwinian Revolution claims man to be an animal deeply rooted in nature, the glory, jest, and riddle of the world. We can read the fossil record and we know that great species of complex animals and plants have disappeared from the face of the earth. We think a simple knowledge of the plagues and famines encountered by mankind seem to be adequate indicators that we court no special care or favors.

I know that evolution-natural selection is a theory and not an observational fact. But this does not make the theory any less important. The theory is not *only* a theory. A theory is a very respectable word in science. There is not the slightest doubt in my mind that man has evolved. I don't think that this means that I could not be convinced otherwise. But I believe I understand how scientific ideas arise and are tested. I can see no really competing scientific theories as to the nature of man. Evolution theory makes immensely good sense to me. But I am aware that each generation of scientists has felt the same about current concepts. My problem is that I sometimes picture myself involved in a debate over politics with the captain of a ship which is heading straight for disaster. Our cries of "One hundred years without Darwin is enough" or "No more pussy-footing about Darwinism" may make some impact on people, but I imagine that it just makes many people go up-tight.

Yet some persons are (in my judgment) deluding themselves by indicating that somehow everything is going to come out just dandy. For many thousands of years man's recent cultural evolution probably has outpaced his biological evolution. One can assume that different cultures have different survival values. Man certainly now adapts his environment to his genes more quickly than his genes are adapted to his environment through natural selection. The question is: can the environmental adaptation proceed quickly enough? It has been said that optimists believe that ours is the best of all possible worlds. (And pessimists are those who fear that the optimists are right.) Some biologists are still optimists, but most are very pessimistic. Perhaps biology will replace economics as the dismal science.

Why the pessimism? The evolutionist understands only too well the nature of reproductive capacity and differential fertility. Even with the exciting discoveries of molecular biology which have elucidated the magic of protein synthesis, we seem quite helpless to combat the dark cloud of overpopulation. The magnificent gains in medicine and genetic counseling seem to fade into irrelevance when compared to the probably stupendous losses which will be associated with the population explosion. It is almost like finding pennies and losing dollars. It is certainly like the absurdity of the bridge on the River Kwai, being built with great care and pride even while the dynamite charge is being placed.

Difficult as it is, I would like to end on a positive note. I think man has the intellectual capacity to solve many of his problems, but he has a greater reproductive capacity to make the problems worse. Things may get better, but only after they get much worse.

Darwin practically opens his book, the *Origin of Species* with a discussion of animals and plants under domestication. Certainly

Darwin assumed that people would recognize the implications of selection going on in front of their eyes. The importance of selection in the controlled evolution of our domestic animals and plants is well known and has had obvious social implications —mostly favorable. The social implications of the development of hybrid corn would be a spectacular example itself.

The deliberate attempts to improve human genetic capacities belong to the area of eugenics. Various degrees of eugenics have been practiced since recorded history so we cannot say the idea was a consequence of evolution theory. There is no doubt that man could start a selective program and produce "breeds" of mankind just as we have done with race horses and pigs. But the complexities of the problems associated with *who* shall be selected and *who* shall do the selecting and to *what end* defy an imaginable solution.

Euthenics, which is environmental engineering, and euphenics, the engineering of human development, seem to offer visions of a brave new world. The medical revolution of transplants, genetic counseling and algeny (the possible insertion of substitute genetic material to cure a genetic defect) are bright spots on the horizon. The difficulty is that there are so many people contemplating the horizon that it is difficult to discern which are the bright spots.

I can see why numbers of people have a difficult time comprehending evolution, which requires an extension of the mind's eye over millions, yes billions, of years because many people cannot even assess the impact of the population pressure which is occurring in front of their eyes. They claim we will be able to produce and distribute enough food even though even now people are starving to death. Of course some of us are not starving. One is reminded of the person in one end of the boat yelling at a person at the other end of the boat that his end of the boat is sinking.

One of Darwin's first basic postulates is that populations,unless controlled, tend to grow beyond the means to maintain them. As a humanistic and civilized society, we are more interested in death control than in birth control, which leads directly to increasing population pressure. But population will indeed be controlled. It probably will be controlled by our old enemies of famine, war, disease, and pestilence. A plague involving a mutant virus might well wipe out millions in a congested area before the clinics could even identify the virus.

If there is any ray of hope it is man's ability to cooperate with his fellows. He is not a lone wolf and desires, in fact requires, a social relationship. Successful cooperation, as well as competition, has important survival value, and cooperation has contributed to man's place as a dominant species among the world's living organisms.

The idea of evolution continues to hold an exciting spot in the academic market place. It has been the focal point of an interesting debate with various audiences being titillated by points scored during the debate. But all of us are members of the same audience and now the debate is coming to a climax. The biologists think they will win the debate but, we hope, without losing the race—human, that is.

LITERATURE CITED

American Medical Association, "Medical Education in the U.S." *Journal of the American Medical Association,* vol. 198, no. 8 (1967) pp. 878, 881.

Bacon, F. in L. W. Taylor, *Physics, The Pioneer Science,* p. 41. Boston: Houghton Mifflin, 1941.

Beach, F. A. "The Snark was a Boojum" (1949). Reprinted in *Readings in Animal Behavior,* T. E. McGill, ed. New York: Holt, Rinehart and Winston, 1965.

Breland, K. and Breland, M. "The Misbehavior of Organisms." *American Psychologist,* vol. 16 (1961), pp. 681-684.

Conant, J. *Science and Common Sense,* p. 10. New Haven: Yale University Press, 1951.

Creation Research Society, *Creation Research Society Quarterly,* W. E. Lammerts, ed., Freedom, California.

Crowe, Beryl L. "The Tragedy of the Commons Revisited." *Science,* vol. 166 (1969), pp. 1103-1107.

DeBakey, Michael E. "Editorial: Human Cardiac Transplantation." *Journal of Thoracic Cardiovascular Surgery,* vol. 55 (1968), pp. 447-451.

DeBakey, Michael E. "Medical Research and the Golden Rule." *Journal of the American Medical Association,* vol. 203 (1968), pp. 574-576.

Disraeli, Benjamin, speech, June 23, 1877.

Dobzhansky, Theodosius. "Changing Man." *Science,* vol. 155 (1967), no. 3761, p. 409.

Fein, Rashi. *The Doctor Shortage: An Economic Diagnosis,* p. 31. Washington, D.C.: The Brookings Institution, 1967.

Fremlin, J. H. "How Many People Can the World Support?" *New Scientist,* no. 415 (1964), pp. 285-287.

Hardin, Garrett. "The Tragedy of the Commons." *Science,* vol. 162 (1968), pp. 1243-1248.

Henderson, Gregory. "Foreign Students: Exchange or Immigration." *International Development Review,* 1964, p. 3 (of reprint).

Hirsch, Jerry (editor). *Behavior—Genetic Analysis,* pp. 1-552. New York: McGraw-Hill, 1967.

Huhn, T. S. *The Structure of Scientific Revolutions,* pp. 2, 5-6, 149-150. Chicago: University of Chicago Press, 1962.

Huxley, Sir Julian. "Evolution in the High-School Curriculum" in *Using Modern Knowledge to Teach Evolution in High School,* p. 26. Chicago: The Graduate School of Education, University of Chicago, 1960.

Jensen, Arthur R. "How Much Can We Boost IQ and Scholastic Achievement?" *Harvard Educational Review,* vol. 39 (1969), pp. 1-123.

Kahn, Herman and Anthony J. Wiener. "Technological Innovation and the Future of Strategic Warfare." *Astronautics and Aeronautics,* vol. 5, no. 2 (1967), pp. 28-48.

Lederberg, Joshua. "Defining the Role of the Scientist as Critic and Reformer." Reprinted from the *Washington Post* in the *Des Moines Register*. September 27, 1969.

Leopold, Aldo. Sand County Almanac. New York: Oxford University Press, 1949.

Lorenz, K. in L. W. Taylor, *Physics, the Pioneer Science,* p. 846. Boston: Houghton Mifflin, 1941.

Morrison, Robert S. "Science and Social Attitudes." *Science,* vol. 165 (1969), pp. 150-156.

Robinson, J. A. T. *Honest to God.* London: S.C.M. Press, Ltd., 1963.

Rogers, Carl R. "Implications of Recent Advances in Prediction and Control of Behavior" in *Contemporary Readings in General Psychology,* ed. Robert S. Daniel, Riverside Press, 1959.

Sigerist, Henry. "The Physician's Profession Through the Ages." *Bulletin of the New York Academy of Medicine,* vol. 9 (1933), p. 661.

Sinsheimer, R. L. "The Prospect for Designed Genetic Change." *American Scientist,* vol. 57 (1969), pp. 134-142.

Skinner, B. F. *Walden Two,* New York: Macmillan, 1948.

Solomon, R. L. "Punishment." *American Psychologist,* vol. 19, (1964), pp. 239-253.

Stoudt, H. W., A. Damon, R. A. McFarland and J. Roberts. "Weight, Height and Selected Body Measurements of Adults, U.S. 1960-62." Publication No. 1000- Series 11 - No. 8 (1965). Washington, D.C.: United States National Center for Health Statistics, Vital and Health Statistics, U.S. Public Health Service-Government Printing Office.

United Nations. "Third Report on the World Helath Situation, 1961-64." *Official Records of WHO,* no 155 (1967), Geneva, p. 35ff.

U.S. Department of Health, Education and Welfare, *Bureau of Health Manpower: An Introduction,* Public Health Service, p. 10.

U.S. Department of Health, Education and Welfare, *Health Manpower Source Book* (physicians, dentists, nurses), Public Health Service, Publication No. 263, Section 9 (1969), pp. 20-22-31.

Weaver, W. "Basic Research and the Common Good." *Saturday Review,* vol. 52, no. 32 (1969), p. 17.

Wiens, John A. "Review of Hirsch, *Behavior-Genetic Analysis.*" *Bioscience,* vol. 18, no. 2 (1968), p. 143.

West, Kelly N. "Foreign Interns and Residents in the U.S." *Journal of Medical Education,* vol. 40 (1965), pp. 1110-1129.

Whitehead, A.N. in I. G. Barbour, *Issues in Science and Religion,* p. 12. Englewood Cliffs, N.J.: Prentice Hall, 1966.

Williams, R. J. "Heredity, Human Understanding, and Civilization." *American Scientist,* vol. 57 (1969), pp. 237-243.